Advances in Intelligent Systems and Computing

Volume 1065

The series "Advances in Intelligent Systems and Computing" contains publications on theory, applications, and design methods of Intelligent Systems and Intelligent Computing. Virtually all disciplines such as engineering, natural sciences, computer and information science, ICT, economics, business, e-commerce, environment, healthcare, life science are covered. The list of topics spans all the areas of modern intelligent systems and computing such as: computational intelligence, soft computing including neural networks, fuzzy systems, evolutionary computing and the fusion of these paradigms, social intelligence, ambient intelligence, computational neuroscience, artificial life, virtual worlds and society, cognitive science and systems, Perception and Vision, DNA and immune based systems, self-organizing and adaptive systems, e-Learning and teaching, human-centered and human-centric computing, recommender systems, intelligent control, robotics and mechatronics including human-machine teaming, knowledge-based paradigms, learning paradigms, machine ethics, intelligent data analysis, knowledge management, intelligent agents, intelligent decision making and support, intelligent network security, trust management, interactive entertainment, Web intelligence and multimedia.

The publications within "Advances in Intelligent Systems and Computing" are primarily proceedings of important conferences, symposia and congresses. They cover significant recent developments in the field, both of a foundational and applicable character. An important characteristic feature of the series is the short publication time and world-wide distribution. This permits a rapid and broad dissemination of research results.

** Indexing: The books of this series are submitted to ISI Proceedings, EI-Compendex, DBLP, SCOPUS, Google Scholar and Springerlink **

More information about this series at http://www.springer.com/series/11156

Mohuya Chakraborty · Satyajit Chakrabarti ·
Valentina E. Balas
Editors

Proceedings of International Ethical Hacking Conference 2019

eHaCON 2019, Kolkata, India

 Springer

Editors
Mohuya Chakraborty
Department of Information Technology
Institute of Engineering and Management
Kolkata, West Bengal, India

Satyajit Chakrabarti
Institute of Engineering and Management
Kolkata, West Bengal, India

Valentina E. Balas
Department of Automation and Applied
Informatics and Department of International
Relations, Programs and Projects, Intelligent
Systems Research Centre
"Aurel Vlaicu" University of Arad
Arad, Romania

ISSN 2194-5357   ISSN 2194-5365   (electronic)
Advances in Intelligent Systems and Computing
ISBN 978-981-15-0360-3   ISBN 978-981-15-0361-0   (eBook)
https://doi.org/10.1007/978-981-15-0361-0

This Springer imprint is published by the registered company Springer Nature Singapore Pte Ltd.
The registered company address is: 152 Beach Road, #21-01/04 Gateway East, Singapore 189721, Singapore

From Editor's Corner

It is with great pleasure that I am serving as the editor for the souvenir of eHaCON 2019 International Ethical Hacking Conference that was held during August 22–25, 2019, at the University of Engineering & Management, Kolkata, India. The ultimate goal for this conference was to create a general awareness of cybercrimes happening in today's world and procedures to countermeasure them. Through a series of keynote talks and research paper presentations on related areas of cybersecurity, ethical hacking, cloud computing, artificial intelligence, modeling and simulation, data analytics, network security, Internet of things, and cryptography, a platform had been created where researchers from India and abroad, from academia and industry, took part in discussion and exchanged their views to make a secured society. The two-day workshop on ethical hacking, gaming competition, and coding competition "De-Cipher" added to the flavor of the overall conference where participants from schools and colleges took part with great interest from India and abroad. Every detail of the conference has been highlighted in the souvenir with the hope that this will become an impetus for new research results in the practical designs of secure systems in the near future.

Dr. Mohuya Chakraborty

Message from Patrons

A very warm welcome to the eHaCON 2019 International Ethical Hacking Conference, which is the first of its series. eHaCON 2019 is an annual event of the Department of Information Technology, Institute of Engineering and Management, Kolkata, India. The main objective of this flagship event is to provide a platform to leading researchers from academia and practitioners from industry in India and abroad to share their innovative ideas, experiences, and cutting-edge research in the areas of cybersecurity, ethical hacking, and network security.

The eHaCON 2019 has been made possible with the generous support of our sponsors: Springer, NASSCOM, Indian School of Ethical Hacking, IEEE IEM ComSoc Student Branch, IEEE IEM CiS Student Branch, CDAC, and HackCieux. I thank all the sponsors, the supporters, and the members of the Department of Information Technology for the grand success of this event.

Satyajit Chakrabarti
President
Institute of Engineering and Management
Kolkata, India
April 2019

Amlan Kusum Nayak
Principal
Institute of Engineering and Management
Kolkata, India
April 2019

Message from Conference General Chairs

It was our great pleasure to extend an affable welcome to all the attendees of eHaCON 2019 International Ethical Hacking Conference organized by the Department of Information Technology, Institute of Engineering and Management (IEM), Kolkata, held at the University of Engineering & Management on August 22–25, 2019. The aim of eHaCON 2019 was to give an open platform where people were able to discuss the implication of new technologies for a secured society. The conference was a balanced mix consisting of technical paper presentations, live demonstrations, workshops, gaming competition, and online coding competition on hacking. The goal was to kick-start the efforts to fully automate cyberdefense. The most substantial new findings of computer network attacks and defenses, commercial security solutions, and pragmatic real-world security experiences were presented in a two-day informative workshop, research paper presentations, and invited talks in the form of a panel discussion on the topic "Present Security Scenario in Digital India" and keynote speeches. Research papers were submitted from ten different countries around the world. Participants in the coding competition were from all over the world.

We express our sincerest thanks to the keynote speakers—Sandeep Sengupta, Indian School of Ethical Hacking; Nirupam Chaudhuri, NASSCOM; Diptiman Dasgupta, IBM; and Atul Agarwal, Apt Software Avenues Pvt. Ltd—for delivering keynote speeches on various cutting-edge topics of security aspects in Digital India.

We are immensely grateful to Maumita Chakraborty for performing an outstanding job for conducting the technical programs. With the help of an excellent committee of international and national experts, very rigorous principles were followed for selecting only the very best technical papers out of a large number of submissions in order to maintain the high quality of the conference.

We would also like to thank Moutushi Singh and Avijit Bose for their outstanding contribution in managing the workshop. Participants were there from various schools, colleges, government offices, and industries. We hope they were immensely benefited from the two-day workshop on ethical hacking.

Our heartiest regards are due to Tapan Kumar Hazra and Arup Kumar Chattopadhyay for creating the Online-Coding-Portal in association with CodeChef and conducting various pre-conference workshops for coding competition "De-Cipher" for the benefit of the participants.

We would like to thank for arranging event participation from various organizations and preparation of call for papers.

Our sincere thanks are due to the Institute of Engineering and Management for co-sponsoring this conference as well as providing both financial and infrastructural supports. We gratefully acknowledge the support of Springer, NASSCOM, IEEE ComSoc Student Branch Chapter of IEM, IEEE Computational Intelligence Student Branch Chapter of IEM, Computer Society of India, Indian School of Ethical Hacking, HackCieux, ITOrizon, CDAC, NASSCOM 10,000 Startups for sponsoring this event, without which the conference could not have been organized on this scale.

We are grateful to all the members of the advisory and technical committees comprising 56 professors and researchers from various parts of the world like Bulgaria, Romania, California, Portugal, UK, Switzerland, Japan, Singapore, and India, for providing their excellent service. We are also thankful to all the local organizing committee comprising Sanchita Ghosh, Baisakhi Das, Lopa Mandal, Pulak Baral (publicity team); Pralay Kar (print team); Sourav Mukherjee (website management); Ankit Anand, Nayan Raj (website development and maintenance); and Kajari Sur, Amit Kumar Mandal, Partha Sarathi Paul, Rabi Narayan Behera, Satyasaran Changdar, Sudipta Paul, and Paramita Mitra (hospitality team) for their hard work and effort to make this conference a grand success.

Last but not least, thanks to all the participants and authors. I hope that they appreciated the conference, and I anticipate that they liked our culturally lively city of joy—Kolkata—as well!

Satyajit Chakrabarti
Director
Institute of Engineering and Management
Kolkata, India
April 2019

Mohuya Chakraborty
Dean
Institute of Engineering and Management
Kolkata, India
April 2019

Message from Organizing Chairs

On behalf of the Organizing Committee of eHaCON 2019 International Ethical Hacking Conference, it was our pleasure to welcome the attendees to the University of Engineering & Management, Kolkata, India.

The conference consisted of four technical sessions with 19 contributed papers, four keynote addresses, two-day workshops along with two-day online coding competition on ethical hacking (De-Cipher) and gaming competition (Capture The Flag). eHaCON 2019 Program Committee, comprising 25 distinguished members, worked hard to organize the technical program. Following the rigorous review process, out of about 100 submissions, only 21 full papers were accepted for presentation in the technical sessions.

Behind every successful event, there lies the hard work, commitment, and dedication of many personalities. Firstly, we wish to thank the entire Program Committee for the excellent job it did in organizing the technical sessions. Special thanks are due to all the reviewers for their obligation in reviewing the papers within a very short time.

We are indebted to the faculty members of the department for managing the two-day workshop on ethical hacking where participants from various schools, colleges, and industries and government officials were benefited. We wish to convey thanks to Swagatam Basu for creating, managing, and conducting the online coding competition "De-Cipher" where more than 220 teams comprising three members per team participated. Pre-conference workshops conducted by the IEEE ComSoc Student Branch Chapter IEM proved to be very successful.

We also thank our collaborators and sponsors like Indian School of Ethical Hacking and HackCieux for the workshop and gaming competition, CodeChef for coding competition, Itorizin, CDAC, NASSCOM, NASSCOM 10,000 startups, IEEE IEM Communication Society Student Chapter, IEEE IEM Computational Intelligence Society Student Chapter, and IEM Computer Society of India Student Chapter for the various other events.

Our sincere gratitude goes to Springer for publishing the conference proceedings in their series "Advances in Intelligence Systems and Computing."

Special thanks go to the conference chair, Mohuya Chakraborty, for giving us immense support and encouragement throughout this period. Once again we hope that all the delegates from India and abroad found the program beneficial and enjoyed the historic city of joy—Kolkata.

We sincerely thank Satyajit Chakrabarti, Director of Institute of Engineering and Management, for his constant support throughout the event.

Last but not least, we thank all our delegates from various industries as well as academia for participation without whom the conference would not have been possible.

Tapan Kumar Hazra
Institute of Engineering and Management
Kolkata, India
April 2019

Moutushi Singh
Institute of Engineering and Management
Kolkata, India
April 2019

Avijit Bose
Institute of Engineering and Management
Kolkata, India
April 2019

Arup Kumar Chattopadhyay
Institute of Engineering and Management
Kolkata, India
April 2019

Program Committee

Patrons

Dr. Satyajit Chakrabarti, President, Institute of Engineering and Management, India
Dr. Amlan Kusum Nayak, Principal, Institute of Engineering and Management, India

General Conference Chairs

Dr. Satyajit Chakrabarti, Director, Institute of Engineering and Management, India
Dr. Mohuya Chakraborty, Institute of Engineering and Management, Kolkata, India

Advisory Committee

Dr. Valentina Emilia Balas, Aurel Vlaicu Univ. of Arad, Romania
Dr. Angappa Gunasekaran, California, State University, Bakersfield, California
Dr. Joao Manuel RS Tavares, University of Porto, Portugal
Dr. Antonio Pescapè, University of Napoli Federico II, Italy
Dr. Ioan Dzitac, Agora University of Oradea, Romania
Dr. Florin Popentiu Vlãdicescu, City University, Northampton Square, London EC1V 0HB
Dr. Robert P. Schumaker, The University of Texas at Tyler
Dr. Sukumar Nandi, IIT Guwahati, India
Dr. Sudipta Roy, Washington University Saint Louis, Missouri, USA
Dr. Ujjwal Maulik, Jadavpur University
Dr. Satish Kumar, 5G Innovation Centre, University of Surrey, Guildford, UK

Dr. Siddhartha Bhattacharyya, VSB Technical University of Ostrava, Czech Republic
Dr. Sangeet Saha, University of Essex, UK
Dr. Partha Dasgupta, Arizona State University
Dr. Peter Boyvalenkov, Institute of Mathematics and Informatics, Bulgarian Academy of Sciences
Dr. Subhash Bhalla, The University of Aizu, Japan
Mr. Sandeep Sengupta, Indian School of Ethical Hacking, India
Dr. Ranjan Mehera, Director of Business Consulting at Subex Inc
Dr. Abhishek Das, Aliah University
Ms. Banani Chakrabarti, IEM
Ms. Gopa Goswami, IEM

Invited Speakers

Mr. Sandeep Sengupta, Indian School of Ethical Hacking, India
Mr. Nirupam Chaudhuri, NASSCOM
Mr. Diptiman Dasgupta, IBM
Mr. Atul Agarwal, APT Softwares Pvt. Ltd.

Editorial Board

Dr. Valentina E. Balas, Professor, "Aurel Vlaicu" University of Arad, Romania
Dr. Mohuya Chakraborty, Institute of Engineering and Management, Kolkata, India
Dr. Satyajit Chakrabarti, Institute of Engineering and Management, Kolkata, India

Technical Program Committee

Dr. Anurag Dasgupta, Valdosta State University, Georgia University, USA
Dr. Yuri Borissov, Institute of Mathematics and Informatics, Bulgarian Academy of Sciences
Dr. Subhash Bhalla, The University of Aizu, Japan
Dr. Subhashis Datta, Bankura University
Dr. S. K. Hafizul Islam, IIIT Kalyani
Dr. G. P. Biswas, IIT (ISM), Dhanbad
Dr. Rik Das, Xavier Institute of Social Service, Jharkhand
Dr. Amlan Chakrabarti, A. K. Choudhury School of Information Technology, University of Calcutta
Dr. Kaushik Deb, CUET, Bangladesh

Dr. Sandip Chakraborty, IIT Kharagpur, India
Mr. Diptiman Dasgupta, Associate Director and Executive IT Architect, IBM
Dr. Karan Singh, JNU Delhi
Dr. Nabendu Chaki, University of Calcutta
Dr. S. Sivasathya, Pondicherry University
Dr. Koushik Ghosh, Burdwan University
Dr. Sourav De, Cooch Behar Government Engineering College
Dr. Parama Bhaumik, Jadavpur University
Dr. Debasish Jana, Professor, Centre for Computer Research, Indian Association for
the Cultivation of Science (IACS)
Dr. Mohuya Chakraborty, IEM
Dr. Amitava Nag, CIT, Kokrajhar
Dr. J. P. Singh, NIT, Patna
Dr. T. Chitralekha, Pondicherry University
Dr. Sudipta Chattopadhyay, Jadavpur University
Dr. Anjan Kumar Kundu, University of Calcutta
Dr. Kaushik Majumder, Maulana Abul Kalam Azad University of Technology

Organizing Committee Chair

Mr. Tapan Kumar Hazra, Institute of Engineering and Management, Kolkata, India
Ms. Moutushi Singh, Institute of Engineering and Management, Kolkata, India
Mr. Avijit Bose, Institute of Engineering and Management, Kolkata, India
Mr. Arup Kumar Chattopadhyay, Institute of Engineering and Management,
Kolkata, India

Organizing Committee

Dr. Satya Saran Changdar, Institute of Engineering and Management, Kolkata,
India
Dr. Amit Kumar Mandal, Institute of Engineering and Management, Kolkata, India
Ms. Maumita Chakraborty, Institute of Engineering and Management, Kolkata,
India
Mr. Partha Sarathi Paul, Institute of Engineering and Management, Kolkata, India
Mr. Rabi Narayan Behera, Institute of Engineering and Management, Kolkata,
India
Mr. Aditya Ray, Institute of Engineering and Management, Kolkata, India
Mr. Partha Bhattacharyya, Institute of Engineering and Management, Kolkata, India
Mr. Animesh Kairi, Institute of Engineering and Management, Kolkata, India
Ms. Paramita Mitra, Institute of Engineering and Management, Kolkata, India
Ms. Kajari Sur, Institute of Engineering and Management, Kolkata, India

Ms. Sudipta Paul, Institute of Engineering and Management, Kolkata, India
Ms. Swagatam Basu, Institute of Engineering and Management, Kolkata, India
Mr. Subindu Saha, Institute of Engineering and Management, Kolkata, India
Mr. Pulak Baral, Institute of Engineering and Management, Kolkata, India

Contents

About the Editors

Mohuya Chakraborty presently holds the post of Dean (Faculty Development) and Professor of the department of Information Technology, Institute of Engineering & Management (IEM), Kolkata. She also holds the post of head of Human Resource Development Centre, IEM. She has done B.Tech and M. Tech from the Institute of Radio Physics and Electronics, Calcutta University in the year 1994 and 2000 respectively and PhD (Engg.) in the field of Mobile Computing from Jadavpur University in 2007. She has successfully completed Chartered Management Institute (CMI Level 5) certification from Dudley College, London, U. K. She is the recipient of prestigious Paresh Lal Dhar Bhowmik Award. She is the member of editorial board of several International journals. She has published 3 patents and over 80 research papers in reputed International journals and conferences. She is the volume editor of Contributed Book; "Proceedings of International Ethical Hacking Conference 2018 eHaCON 2018", in Springer series "Advances in Intelligent Systems and Computing book series (AISC volume 811)", published by Springer. She has handled many research projects funded by the DST, AICTE, CSIR and NRDC, and has published a number of papers in high-impact journals. Her research areas include network security, cognitive radio, brain computer interface, parallel computing etc. She is a member of IEEE Communication Society and IEEE Computational Intelligence Society as well as the faculty adviser of IEEE Communication Society and IEEE Computational Intelligence Student Branch Chapters of IEM, Kolkata Section.

Satyajit Chakrabarti is Pro-Vice Chancellor, University of Engineering & Management, Kolkata and Jaipur Campus, India and Director of Institute of Engineering & Management, IEM. As the Director of one of the most reputed organizations in Engineering & Management in Eastern India, he launched a PGDM Programme to run AICTE approved Management courses, Toppers Academy to train students for certificate courses, and Software Development in the field of ERP solutions. Dr. Chakrabarti was Project Manager in TELUS, Vancouver, Canada from February 2006 to September 2009, where he was intensively involved in planning, execution, monitoring, communicating with stakeholders, negotiating with vendors and cross-functional teams, and motivating members. He managed a team of 50 employees and projects with a combined budget of $3 million.

Valentina E. Balas is currently Full Professor in the Department of Automatics and Applied Software at the Faculty of Engineering, "Aurel Vlaicu" University of Arad, Romania. She holds a Ph.D. in Applied Electronics and Telecommunications from Polytechnic University of Timisoara. Dr. Balas is author of more than 300 research papers in refereed journals and International Conferences. Her research interests are in Intelligent Systems, Fuzzy Control, Soft Computing, Smart Sensors, Information Fusion, Modeling and Simulation. She is the Editor-in Chief to *International Journal of Advanced Intelligence Paradigms (IJAIP) and to International Journal of Computational Systems Engineering (IJCSysE)*, member in Editorial Board member of several national and international journals and is evaluator expert for national, international projects and PhD Thesis. Dr. Balas is the director of Intelligent Systems Research Centre in Aurel Vlaicu University of Arad and Director of the Department of International Relations, Programs and Projects in the same university. She served as General Chair of the International Workshop Soft Computing and Applications (SOFA) in eight editions 2005-2018 held in Romania and Hungary. Dr. Balas participated in many international conferences as Organizer, Honorary Chair, Session Chair and member in Steering, Advisory or International Program Committees. She is a member of EUSFLAT, SIAM and a Senior Member IEEE, member in TC – Fuzzy Systems (IEEE CIS), member in TC - Emergent Technologies (IEEE CIS), member in TC – Soft Computing (IEEE SMCS). Dr. Balas was past Vice-president (Awards) of IFSA International Fuzzy Systems Association Council (2013-2015) and is a Joint Secretary of the Governing Council of Forum for Interdisciplinary Mathematics (FIM), - A Multidisciplinary Academic Body, India.

Image Processing

Review and Comparison of Face Detection Techniques

Sudipto Kumar Mondal, Indraneel Mukhopadhyay and Supreme Dutta

Abstract Automatic object detection is a common phenomenon today. To detect an object first thing is captured, is the image of the object. Now in an image categorically different types of objects are possible. Here, we are considering human face as a most common object. Day by day, the number of application based on face detection is increasing. So the demand of highly accurate and efficient face detection algorithm is on the high. In this paper, our motive is to study different types of face detection techniques and compare them. Various face detection techniques like using Haar-like cascade classifier, Local Binary Pattern cascade classifier and Support Vector Machine-based face detection methods are compared here. All these techniques are compared based on time, accuracy, low light effect, people with black face and with false object and based on memory requirement.

Keywords Face detection · Haar cascade classifier · Local Binary Pattern cascade classifier · Support Vector Machine-based face detection

1 Introduction

Face detection is the process of finding faces in an image or in a frame and, if present, return the location [1] and according to that it is marked. It is generally used in detecting a human face. It also detects eyes of a human in an image or in a frame. Many techniques are there for detecting faces [2, 3]. Steps for face detection are

S. K. Mondal (✉) · S. Dutta
University of Engineering & Management, Kolkata, India
e-mail: sudipto.mondal@uem.edu.in

S. Dutta
e-mail: supremedatta@gmail.com

I. Mukhopadhyay
Institute of Engineering & Management, Kolkata, India
e-mail: imukhopadhyay@gmail.com

© Springer Nature Singapore Pte Ltd. 2020
M. Chakraborty et al. (eds.), *Proceedings of International Ethical Hacking Conference 2019*, Advances in Intelligent Systems and Computing 1065, https://doi.org/10.1007/978-981-15-0361-0_1

- Importation of an image [4].
- Transformation of the image from RGB to grayscale.
- Division of objects in the image to acquire quick detection.
- Use classifier which is used to detect faces in the image.
- Make rectangle box in the image by coordinating x, y, w, h.

2   Applications of Face Detection

There are several applications based on face available around us in real life. These applications boost the overall performance.

Camera autofocus: When the camera takes photographs, it detects people's faces and helps the camera to autofocus. This approach ultimately helps in taking good photographs.
Facebook: With respect to face, automatic tagging is done there.

- Face Recognition: The most endemic use of face detection is face recognition. Face recognition is used to identify people's faces. This is used in apps, phones, airports, companies, etc. Nowadays, it is also used in the attendance system.
  OpenCV (Open source computer vision) is a cross-platform repository of program which helps us in real-time computer vision. There are three built-in face recognizers in OpenCV.

  There are three built-in face recognizers in OpenCV:

- Eigenfaces Face Recognizer [5, 6]
- Fisherfaces Face Recognizer [7]
- Local Binary Patterns Histograms Face Recognizer [8].

3   Face Detection Techniques

There are several face detection techniques [9] available. Following three face detection techniques are discussed here.

3.1   Face Detection Using Haar Cascade Classifier

Haar cascade classifier is proposed by Viola–Jones face detection algorithm [10]. This algorithm needs a lot of positive images and negative images to train the classifier. Positive images are those which have faces in the image and negative images are those which do not have faces in the image [7]. After this, the features are extracted. There are some features in Haar cascade classifier

- Edge feature
- Line feature
- Four rectangle feature.

We need to subtract the sum of pixels under the white rectangle from the sum of pixels under black rectangle to get the feature. But it needs a lot of calculation ($24 \times 24 = 160,000 +$ features). So we need to find the best features which are achieved by AdaBoost [3, 11]. The features are reduced to 6000. But we know that in an image non-face image region is more than face region. So we do not need to focus on the non-face region, and the features are grouped into different stages. For example, if a window is not able to pass the first stage then we do not need to apply the next stages further. We will discard it. But if a window passes the first stage then we will apply the second stage to it and continue, and if a window passes all stages then this is a face region. This is how Viola–Jones face detection works.

3.2 Face Detection Using Local Binary Pattern Cascade Classifier

The image of a face is divided into texture descriptor and local regions. This is done independently on each region. The descriptors are concatenated to form the global descriptor.

This is a face detection cascade classifier where the training image is divided into blocks. It recognizes every image as full of micro-patterns. Due to that, local facial component is extracted first using LBP [11] operator. This operator extracts the texture from the face in terms of a binary pattern. Pattern can be of flat areas, curved edge, spots, flat areas, etc. Then, it generates a histogram from that. From every sub-region, LBP histogram are extracted and combined into a single, spatially enhanced histogram.

Face detection using Local Binary Pattern depends mainly on the appearance of faces. That means more and more presence of positive images makes this method more effective.

Local Binary Pattern mainly follows the followings:

- The LBP identifies histogram from the patterns on a pixel level.
- The labels are integrated to generate pattern of regions.
- The regional histograms are concatenated to build a global description of the face.

3.3 Support Vector Machine

SVM [12] can be used to make a binary classification. SVM finds a hyper-plane (line in 2D, plane in 3D, etc.). It separates its training data in such a way that the distance between the hyper-plane and the closest points from each class is maximized once SVM finds this hyper-plane, you can classify new data points by seeing which side of this hyper-plane landing on SVM can only be used on data that are linearly separable (i.e., a hyper-plane can be drawn between the two groups. However, it is not so, as a common way to make data separable in a linear way is to map it to a higher size (but be careful, as this is computationally expensive). You can map it as you wish, but there are ways to do this, they are called Kerns. When using a combination of these Kernels and modifying their parameters, you'll most likely get better results than doing it your way. The really interesting thing about the SVMs is that it can or use them when you have very little data compared to the number of functions of each of your data points, in other words when the number of data for the number of features per data ratio is low. Normally when this ratio is low, over-processing occurs, but since the SVMs only use some of your data points to create the hyper-plane, in the first place, it does not really matter that you give it such small data. However, note that accuracy of forecasts is reduced when very little data is used. The SVMs simply tell you in which class a new data point falls, not the probability that it is in that class. This is obviously a disadvantage.

4 Comparative Evaluation

Above specified face detection techniques are compared by calculating the time that is how much time a technique is taking to detect faces in an image. All the techniques are run in PYTHON 3.7.2 using OpenCV 4.0.0.21 [2]. Cascade classifiers are pre-trained to detect faces and other objects where Haar cascade classifier is pre-trained using Viola–Jones face detection algorithm.

4.1 Single Face Detection

Haar cascade classifier took 0.399 s to detect the face in Fig. 1a. Local Binary Pattern cascade classifier took 0.368 s to detect the face in Fig. 1b.

Real-time face detection using Support Vector Machines took 1.84 s to detect the face as shown in Fig. 1c.

So, we can say that, among all these three techniques, real-time face detection using Local Binary Pattern cascade classifier takes minimum time to detect the face, and Support Vector Machine takes the maximum time.

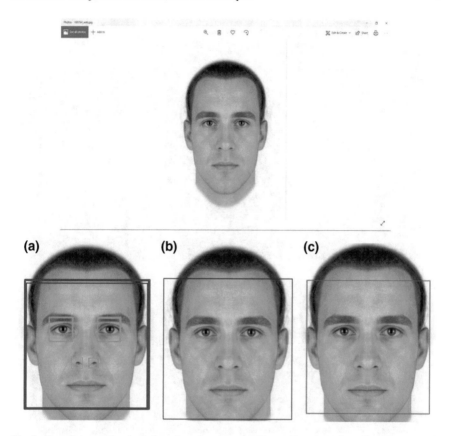

Fig. 1 Sample image for single face detection time checking. **a** Face detection using Haar cascade classifier. **b** *Face* detection using Local Binary Pattern cascade classifier. **c** Real-time face detection using Support Vector Machines

4.2 Accuracy Detection

Next comparison is based on accuracy, that is, how much a technique is accurate to detect faces in an image.

Haar cascade classifier detected 18 faces as shown in Fig. 2a. It took 0.901 s to detect the faces as shown in Fig. 2a. Local Binary Pattern cascade classifier took 0.526 s and detected 14 faces in as shown in Fig. 2b. Support Vector Machines took 2.35 s and detected 17 faces in as shown in Fig. 2c.

So, we can say that, among all these three techniques, real-time face detection using Local Binary Pattern cascade classifier takes minimum time to detect the faces, and Support Vector Machine takes the maximum time and the most accurate face detection method is Haar cascade classifier which has a high detection rate and the least accurate being the Local Binary Pattern cascade classifier which has a low detection rate.

Fig. 2 Sample image for accuracy checking for multiple face detection. **a** Face detection using Haar cascade classifier. **b** *Face* detection using Local Binary Pattern cascade classifier. **c** Real-time face detection using Support Vector Machines

4.3 Black Face Detection

Now, we checked that how these techniques work on black face.

Haar cascade classifier did not detect the face as shown in Fig. 3a.

Local Binary Pattern cascade classifier detected the face as shown in Fig. 3b.

Real-time face detection using Support Vector Machines did not detect the face as shown in Fig. 3c.

So, we can say that, among all these three techniques, real-time face detection using Local Binary Pattern cascade classifier can detect faces with very dark complexion and Support Vector Machines and Haar cascade classifier cannot detect faces with very dark complexion.

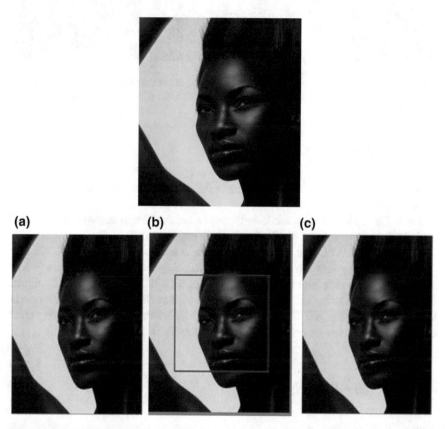

Fig. 3 Sample Image for black face detection. **a** Face detection using Haar cascade classifier. **b** *Face detection using Local Binary Pattern cascade classifier.* **c** Real-time face detection using Support Vector Machines

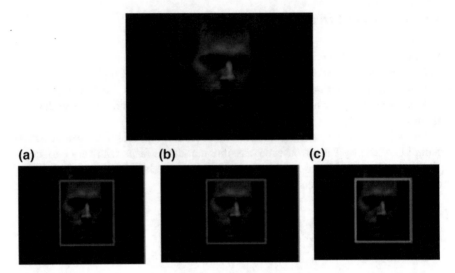

Fig. 4 Sample image to detect face in low light. **a** Face detection using Haar cascade classifier. **b** *Face* detection using Local Binary Pattern cascade classifier. **c** Real-time face detection using Support Vector Machines

4.4 Low Light Effect

Haar cascade classifier detected the face in 0.332 s but could not detect the eyes as shown in Fig. 4a.

Local Binary Pattern cascade classifier detected the face as shown in Fig. 4b and took 0.3261 s to detect the face in low light effect.

Support Vector Machines has detected the face as shown in Fig. 4c in 1.798 s.

So, we can say that the face detection using Support Vector Machines is less accurate in low light effect because it takes highest time to detect in low light, and the face detection using Local Binary Pattern cascade classifier is most accurate in low light effect because it takes the least time. In real-time face detection using Haar cascade classifier, the accuracy is moderate in low light effect.

4.5 Non-face Detection

Figure 5a is an image which looks like human face. Haar cascade classifier has identified it as human face.

Local Binary Pattern has identified that there is no human faces as shown in Fig. 5b.

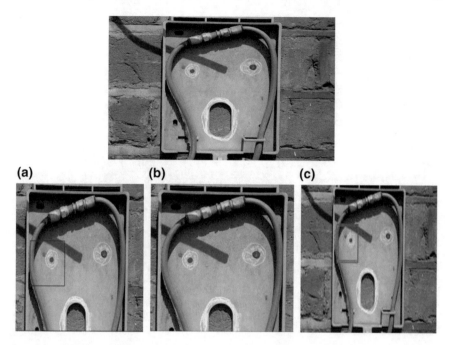

Fig. 5 Sample image to detect non-face. **a** Face detection using Haar cascade classifier. **b** *Face detection using Local Binary Pattern cascade classifier.* **c** Real-time face detection using Support Vector Machines

Figure 5c is an image which looks like human face. Support Vector Machine has identified it as human face.

So, we can conclude that the false positive rate is high in face detection using Haar cascade classifier and Support Vector Machine. False positive rate is low in face detection using Local Binary Pattern cascade classifier.

4.6 Face with Obstacles

Haar cascade classifier has detected the face with spectacles and beard as shown in Fig. 6a. Local Binary Pattern cascade classifier did not detect the face with beard and spectacles as shown in Fig. 6b.

Support Vector Machine detected the face with obstacles as shown in Fig. 6c.

So, we can say that the face detection in case of face with obstacles using Local Binary Pattern cascade classifier is ineffective comparing to other two method as the have detected effectively.

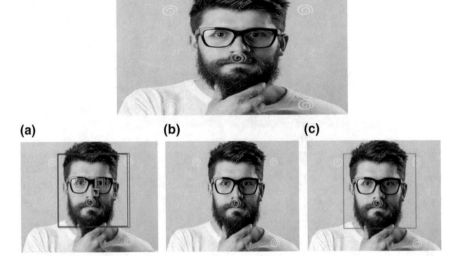

Fig. 6 Sample image to check face with obstacles. **a** Face detection using Haar cascade classifier. **b** *Face* detection using Local Binary Pattern cascade classifier. **c** Real-time face detection using Support Vector Machines

Comparative analysis

Approach	Haar	LBP	SVM
Average time (single face)	0.43004 s	0.42646 s	2.30883 s
Accuracy	High	Low	High
Black face detection	No	Yes	No
Low light effect	Moderate accuracy	Best accuracy	Least detection
Face with obstacles	High detection	No detection	High detection
Object similar to face	High	Low	High
System usage	CPU-71% Memory-45%	CPU-89% Memory-43%	CPU-66% Memory-44%
Time (multiple faces)	1.04261 s	0.67998 s	3.07367 s

Proposed Algorithm

From the above analysis, we are observing that Haar cascade classifier's overall performance is good. In some cases, it is not working efficiently specially in case of black face detection or in case of low light images. In low light, Haar method has detected the face but unable to detect the eye. Even if in case of face with obstacles the performance is poor of Haar cascade classifier. So from the above observation, we are proposing an algorithm where all the feature of Haar cascade classifier must be there and along with that some more feature to resolve the above-mentioned problems are must be there.

5 Conclusion

We are surrounded by various applications based on face detection today like using face as a mobile password to automated door opening using face. Various techniques are available to detect faces as well as recognize them. In this paper, face detection using Haar cascade classifier and Local Binary Pattern cascade classifier and real-time face detection using Support Vector Machines has been compared and analyzed based on certain test cases which have not been considered earlier. From that analysis, the following observations are coming:

Face detection using Haar cascade's performance is good in many cases and poor in case of dark complexion or low light and in case of false image.

Those areas where Haar cascade classifier is weak, Local Binary Pattern cascade classifier works efficiently there. But this algorithm having the problem of accuracy and in case of face having some obstacles its performance is poor. Another important drawback of linear binary pattern approach is consumption of maximum CPU during run time. With respect to other two algorithm, real-time face detection using Support Vector

Machine has no significant advantages.

References

1. Chauhan, M., Sakle, M.: Study & analysis of different face detection techniques. Int. J. Comput. Sci. Inf. Technol. (IJCSIT) 5(2), 1615–1618 (2014). (Parul Institute of Engineering & Technology)
2. Zhang, C., Zhang, Z.: A Survey of Recent Advances in Face Detection. Microsoft Research Microsoft Corporation, June 2010
3. Jones, M., Viola, P.: Fast Multi-view Face Detection. Mitsubishi Electric Research Lab TR-20003-96 (2003)
4. Singh, Y.K., Hruaia, V.: Detecting face region in binary images. In: 2015 IEEE Recent Advances in Intelligent Computational Systems (RAICS)
5. Kshirsagar, V.P., Baviskar, M.R., Gaikwad, M.E.: Face recognition using eigenfaces. In: 3rd IEEE International Conference on Computer Research and Development (2011)
6. Kshirsagar, V.P., Baviskar, M.R., Gaikwad, M.E.: Face recognition using eigenfaces. In: 2011 3rd International Conference on Computer Research and Development
7. Lee, H.-J., Lee, W.-S., Chung, J.-H.: Face recognition using Fisherface algorithm and elastic graph matching. In: Proceedings 2001 IEEE International Conference on Image Processing (2001)
8. Taouche, C., Batouche, M.C., Chemachema, M., Taleb-Ahmed, A., Berkane, M.: New face recognition method based on local binary pattern histogram. In: 15th IEEE International Conference on Sciences and Techniques of Automatic Control and Computer Engineering (STA) (2014)
9. Reney, D., Tripathi, N.: An efficient method to face and emotion detection. In: 2015 Fifth International Conference on Communication Systems and Network Technologies
10. Paul, V., Jones, M.J.: Rapid object detection using a boosted cascade of simple features. In: Proceedings of the 2001 IEEE Computer Society Conference on Computer Vision and Pattern Recognition, 2001. Vol. 1, pp. 511–518 (2001)

11. Jo, C.: Face detection using LBP features. CS 229 Final Project Report (2008)
12. Shavers, C., Li, R., Lebby, G.: An SVM-based approach to face detection. In: 2006 Proceeding of the Thirty-Eighth Southeastern Symposium on System Theory

Keyword Spotting with Neural Networks Used for Image Classification

Ashutosh Kumar, Sidhant Mishra and Tapan Kumar Hazra

Abstract We might be living in a Screen Age, almost everyday a new object with a bright touch screen is invented. A possible antidote to our screen addiction is voice interface. The available voice assistants are activated by keywords such as "hey Siri" or "okay Google" [1]. For initial detection of these keywords, it is impractical to send the audio data over the Web from all devices all the time, as it would increase the privacy risks and would be costly to maintain. So, voice interfaces run a keyword detection module locally on the device. For independent makers and entrepreneurs, it is hard to build a simple speech detector using free, open data, and code. We have published the result as easy to train "Kaggle notebooks" [2]. With considerable improvement, these models can be used as a substitute for our keypads in touch screens. In this work, we have used convolutional neural networks (CNNs) for detection of the keywords, because of their ability to extract important features, while discarding the unimportant ones. This results in smaller number of parameters for the CNNs as compared to the networks with fully connected layers. The network that we have used on this work is derived from the CNNs that gave state-of-the-art results for image classification, e.g., dense convolutional network (DenseNet) [3], residual learning network (ResNet) [4], squeeze-and-excitation network (SeNet) [5], and VGG [6]. We have discussed the performance of these CNN architectures for keyword recognition. The method for reproducing the result had been suggested as well. These models achieve top one error of ~96–97%, with the ensemble of all achieving ~98%, on the voice command dataset [7]. We have concluded by analyzing the performance of all the ten models and their ensemble. Our models recognize some keywords that were not recognized by human. To promote further research (https://github.com/xiaozhouwang/tensorflow_speech_recognition_solution) contains the code.

A. Kumar (✉) · S. Mishra · T. K. Hazra
Institute of Engineering & Management, Salt Lake, Kolkata, India
e-mail: aks2k17@gmail.com

S. Mishra
e-mail: sidhant.plb@gmail.com

T. K. Hazra
e-mail: tapankumar.hazra@iemcal.com

© Springer Nature Singapore Pte Ltd. 2020
M. Chakraborty et al. (eds.), *Proceedings of International Ethical Hacking Conference 2019*, Advances in Intelligent Systems and Computing 1065, https://doi.org/10.1007/978-981-15-0361-0_2

Keywords Voice command · Keyword recognition · ResNet · DenseNet · SeNet · VGG

1 Introduction

1.1 The Problem Statement

In this work, we propose algorithms to detect keywords in voice commands. This algorithm, when implemented, could improve the voice assistants [8] or help us to get rid of the touch-enabled keypads. Voice assistants continuously listen for specific words such as "Hey, Siri," "Okay Google" [1] and "Hey, Cortana," respectively, to initiate interaction. It is not practically possible to send all the voice data over the Web to remote servers as this would increase the privacy risk for the users and the data transfer cost for the organizations.

1.2 Current Methodology

The keyword recognizing system at Google, which uses deep neural networks (DNNs) have performed better than hidden Markov models [1]. DNNs have small number of parameters, and this makes them ideal for resource-constrained devices. CNNs have not only shown the improvement in keyword recognition task but they have also outperformed other models for other acoustic modeling tasks [9–13].

Voice has got a strong correlation in time and frequency, and CNNs help us to model these correlations by sharing the weights across local regions of the input space. DNNs ignore such correlations, whereas weight sharing has shown good performance in other fields [14]. When the speaking style changes, this shifts the formants in frequency domains. It is hard for DNNs to model such variance, whereas the CNNs can model these variances in time and frequency domain by averaging the output of different layers. CNNs can also extract the features that will reduce such variances. CNNs have also shown an increase of 27% in false reject (FR) when they stride in frequency, when they pool in time an improvement of 41% is seen, and an improvement of 6% is seen when they pool in frequency only.

ImageNet Large-Scale Visual Recognition Challenge (ILSVRC) [15] has played an important role in the development of deep neural networks. Some of the best models that were developed used high-dimensional feature encoding [16]. Some models used small filters and small strides [17, 18] in first convolution layer, whereas some trained the model over whole training and test dataset and over different sizes [19]. Another approach to improve the accuracy of model is by increasing the depth of the models. CNNs extract and combine features from different layers [17], some of the well-performing models [20–23] that were trained using challenging datasets [15] have depth varying from sixteen [22] to thirty [21]. Deep models have been used for many other visual recognition tasks [6, 24–27] as well. Increasing in depth of models has resulted in either vanishing or exploding gradients [28, 29]. Some of the methods to handle exploding and vanishing gradient are by normalized initialization [23, 29–31] and adding normalization to the intermediate layers [21], and this enables converging of stochastic gradient descent (SGD) with backpropagation [32]. As we increase the depth of models, the accuracy tends to decrease, which is not due to over-fitting, but it is due to the fact that adding more layers to a suitably deep network increases the training error [33, 34]. In a residual learning framework, the training of deep networks is done by learning the residual function for the input layer instead of learning an unreferenced function. Identity connection helps the deep models such as LeNets [35], VGG [15], highway networks [33], and residual networks [4] to bypass signals from one layer to the other.

In a DenseNet [3], if there are L layers, then the total number of connections will be $L*(L + 1)/2$. This is to ensure that all the layers of DenseNet [3] are connected to each other.

DenseNet [3] helps to get rid of the vanishing gradient problem, strengthens feature propagation, reuses features, and reduces the number of features. The "squeeze-and-excitation" (SE) block [5] models the interdependencies between the channels; when we stack these SE blocks, we get the SeNet, which generalizes very well across different datasets. These architectures also reduce the computation required for training the models. The spatial correlation between features is also integrated by inception family of architectures [20, 21].

In the presented work, we have used the network architectures [3–5, 22] developed for the image classification task and used it to classify the keywords in voice commands [36]. We have used Mel spectrogram and MFCC for preprocessing the audio clips and then reshaped the input for the models. The different models achieved a score of ~96% (Fig. 1) and the ensemble of all achieving 98.01%.

1.3 Related Works

Thomas O'Malley's work [37] uses 120 log mel filter bank. The models treated time and frequency differently. The model architecture is used, and its interpretation is shown in Fig. 2.

The other two techniques that worked are

1. Standardized peak (windowed) volume, every clip was split into 20–50 chunks, and then the volume of the clip was standardized so that every clip had the same maximum chunk volume.
2. Vocal tract length perturbation as described in [38].

Fig. 1 Accuracy of different models

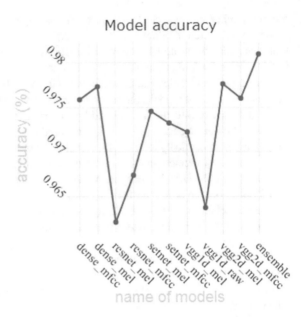

2   Approach

2.1   Preprocessing

We load the audio file as floating-point time-series data and resample it to the rate of 16,000 and if the sampled audio is smaller in size, we pad them with constant values (zeros). And if the sampled audio file is larger than 16,000, we truncate it. During training, we add noise to the clip, while during testing we do not add any noise to the voice clips. While adding the noise, we randomly draw a "multiplying factor" from the uniform distribution in (0.75, 1.25), a random number from {1,2} as the num_noise and then a random number from {0.1, 0.5, 1, 1.5} as max_ratio. We chose a random number from [0.1, 0.3] as mix_noise_probability and also a random integer between {80 and 120} as the measure for the shift range. We perform the time shift on the loaded audio, then add noise to it, and scale it by a factor of "multiplying factor." Once we have augmented the original audio data, we calculate its mel-scaled spectrogram, with sample rate = 16000, with number of mel bins = 40, number of samples between successive frames = 160, and fast_fourier_window of 480 (and the minimum frequency of 20 and the maximum frequency of 4000).

Fig. 2 Interpretation of all the eight layers are as follows: (i) "denoising" and basic feature extraction layer; (ii) getting back down to standard 40 frequency features; (iii) looking for local patterns across frequency bands; (iv) allows for speaker variation similar to the one proposed in [2]; (v) this layer treats each remaining frequency band differently and compresses the frequency dimension entirely, and it detects the phoneme-level features; (vi) it looks for the connected component of a short keyword at different points in time; (vii) collects all the components

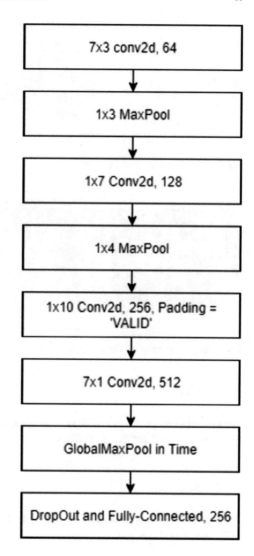

2.1.1 MFCC

For MFCC, we consider the spectrogram whose values are greater than 0 and then perform log transformation them. Then we create a discrete cosine transformation (DST-II) filter with number of input bins as 40 and the number of input filters as 40.

We take every column vector of the spectrogram and perform matrix multiplication with the 40×40 DCT filter, and this gives us 101, 40×1 vectors which we latter stack to get the required MFCC of shape (40×101).

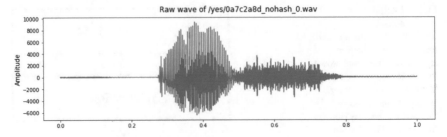

Fig. 3 Raw wave of a sound clip "yes"

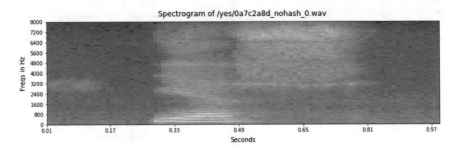

Fig. 4 Spectrogram of a voice clip "yes"

2.1.2 Mel

We convert the power spectrogram (amplitude squared) to decibel (dB) units. If required, we perform the normalization.

2.1.3 Raw

We have also simply reshaped the audio data and used it to train our models, after normalization (Figs. 3, 4, 5, and 6).

3 Network Architectures and Configurations

3.1 VGG-Like Model

The models that we have developed, motivated by VGG, are shown in Figs. 7, 8, and 9.

The model architecture shown in Fig. 7 is called vggmel, as the input fed into this after extracting the features using mel. The input after feature extraction, size

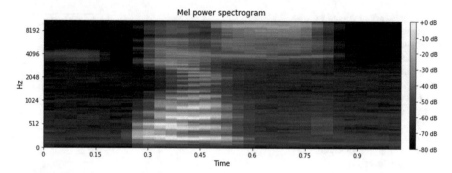

Fig. 5 Mel power spectrogram of a sound clip "yes"

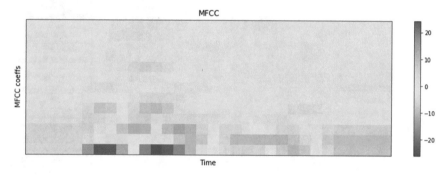

Fig. 6 MFCC spectrogram of a sound clip "yes"

(40 × 101), is fed to the stack of convolutional layers as shown in Fig. 7. The first 20 layers shown in Fig. 7 have a combination of "1D convolutional layers," "batch normalization layers," and max pooling layers. The size of the convolving kernel is 3, with the stride of convolution as 1 and the padding as 1. The kernel size for max pooling is 2 with a stride of 1. After these stacks of feature extracting layers, the 1D input of 1536 is fed to the classifier layer, last 5 layers, which give us the predicted probability corresponding to each label. The classifier layers consist of 3 linear layers, and these linear layers use the ReLU activation unit and the dropout with probability 50%. This model achieves the accuracy of 97.22%. The training and validation error are shown in Fig. 10.

The model shown in Fig. 8 is vgg2d, so-called because the convolutional kernels used in this are two-dimensional in shape. The feature of audio is extracted using the mel (Sect. 2.1). The size of input fed into the architecture is (1 × 128 × 128). The input is fed to the feature extraction stack of 31 layers, which comprises of the convolutional layer with two-dimensional convolutional kernel (3 × 3), stride of 1 and padding of 1, max pooling layer with kernel size 2 × 2 and stride of 1, and the 2d-batch normalization layer with ReLU activation. After the feature extraction stack, the data of size 256 × 4 × 4 is passed through adaptive_max_pooling_2d layer to yield

22

A. Kumar et al.

Fig. 7 Network architecture, the convention followed for convolutional layer, Conv1donv1d (in_channel, out_channel, kernel_size, padding, stride), for the batch normalization layer the convention followed is [conv1d, BatchNorm1d (number_channel), ReLU], and the convention followed for the MaxPool1d is MaxPool1d (kernel_size, stride)

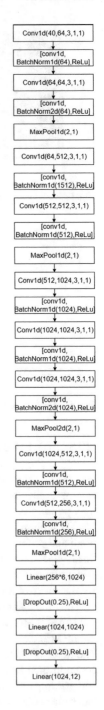

Fig. 8 VGG2d architecture, the convention used for the convolutional layers is Conv2d (in_channels, out_channel, kernel_size, padding), for the batch normalization layer, the convention followed is [conv2d, BatchNorm2d (number_channel), ReLU], and for the max pooling layer, the convention followed is MaxPool2d (kernel_size, stride). The stride of 1 is not shown for the convolutional layers for brevity

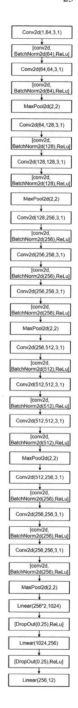

Fig. 9 VGG-like
architecture, the convention
followed for convolutional
layer, Conv1d (in_channel,
out_channel, kernel_size,
padding, stride), for the
batch normalization layer the
convention followed is
[conv1d, Batch-
Norm1d(number_channel),
ReLU], and the convention
followed for the MaxPool1d
is MaxPool1d(kernel_size,
stride)

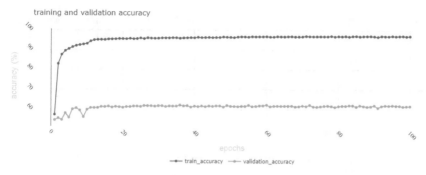

Fig. 10 Plot of training accuracy versus validation accuracy for VGG1d_mel

data of size 256 × 1 × 1. The output of adaptive_max_pooling_2d layer is fed to the adaptive_average_pooling layer, which does not change the size of the data. Further, the output of the adaptive_average_pooling layer and the adaptive_max_pooling_2d layer are concatenated to give a feature vector of size 521. This feature vector of size 512 is fed to the final classifier (last 5 layers in Fig. 8), comprising of 3 linear layers, 2 dropout layers with probability 25% and ReLU as the activation function to give us the predicted probability of all the 12 labels. The model achieves an accuracy of 97.76%. Vgg2d model shown in Fig. 8 is also used to classify the keywords in speech command by the use of MFCC (Sect. 2.1). Following the above configurations, an accuracy of 97.60% was achieved.

VGG1d model shown in Fig. 9 uses the raw features extracted (Sect. 2.1). The input to the feature extraction stack (first 33 layers in Fig. 17) is of size (1 × 16,000). After going through different convolutional layers with one-dimensional convolutional kernels, one-dimensional batch normalization layer and max pooling layer, the output of the feature extraction layer is of shape (256 × 31). Further after the (256 × 31), the input is passed through the adaptive_max_pool_1d layer and we get feature vector of size (256 × 1). This feature vector of size (256 × 1) is further fed into the adaptive_average_pool_1d layer which leaves the feature vectors unaltered. Finally, the output of adaptive_max_pool_1d layer and adaptive_average_pool_1d layers is concatenated to provide the input vector of (512 × 1) to the classifiers. The classifiers comprising of 3 linear layers, 2 dropout layers with probability 50% and ReLU activation unit gives us the predicted probability for different labels. The accuracy achieved by this model and the preprocessing method (Sect. 2.1.3) is 96.37%.

3.2   ResNet-Like Models

Model shown in Fig. 15 is derived from the architecture of ResNet, and there are two different feature extraction techniques that are used to give two different models (a) ResNet_mel, that uses the feature extraction technique described in Sect. 2.1 and (b) ResNet_mfcc, that uses the feature extraction technique described in Sect. 2.1.

To best describe the forward function of the ResNet (mel/mfcc) model, we present the forward function used to train the model (Figs. 11, 12, 13, and 14).

```
For i in range (self.n_layers + 1):
    y = F.relu(getattr (self, "conv{}".format(i)) (x))
    if i == 0:
        old_x = y
```

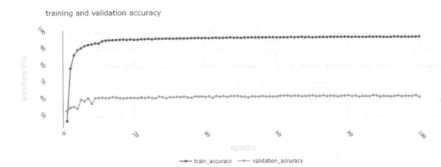

Fig. 11 Plot of training accuracy versus validation accuracy for VGG1d_raw

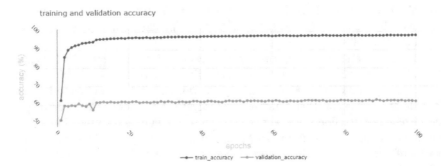

Fig. 12 Plot of training accuracy versus validation accuracy for vgg2d_mfcc

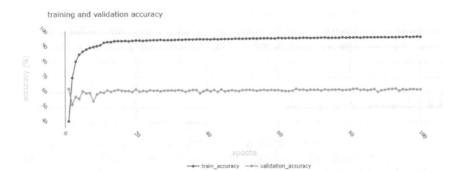

Fig. 13 Plot of training accuracy versus validation accuracy for vgg2d_mel

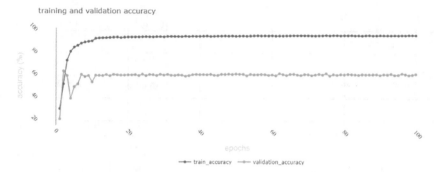

Fig. 14 Plot of validation accuracy versus training accuracy for ResNet_mfcc

```
if i > 0 and i % 2 == 0:
    x = y + old_x
    old_x = x
else:
    x = y
if i >0:
    x = getattr(self, "bn{}".format(i))(x)
x = x.view(x.size(0), x.size(1), -1)
x = torch.mean(x, 2)
```

The name of layers can be referenced from Fig. 15. The size of input to layer 0 is (40 × 101) and so is the size of input and output to all the layers. It should be noted that input–output from the even layer are added to the alternate even layers, and then this trend follows as shown in the code snippet above. It should be noted that though the block diagram in Fig. 15 shows Conv_i followed by bn_i (where 0 < i < 10), the forward function used it in reverse order that is bn_i followed by Conv_i. After all the convolutional layers and the batch normalization layers, we get an input of size 128 × 40 × 101. This input is further resized to 128 × 4040, which when passed through the last (linear) layer gives us the probability of all the 12 labels for given input. The accuracy achieved when using the mel, feature extraction technique is 96.21%, and when using the MFCC feature extraction technique, the accuracy achieved is 96.73%.

3.3 DenseNet

In the proposed DenseNet, the first stack is of a two-dimensional convolutional layer with kernel size 7 × 7, stride of 2 and padding of 3 that accepts an input of size (128 × 128), and it followed by a two-dimensional batch normalization layer, a ReLU activation unit and finally a MaxPool2d layer with kernel_size = 3, stride = 2 and padding = 1. The first stack is followed by number of dense blocks, and a dense block is formed of dense layers followed by transition layer (Figs. 16, 17, and 18).

Fig. 15 Block diagram of model inspired by the ResNet. The model shows combination of convolutional layers and batch normalization layers placed in alternate position

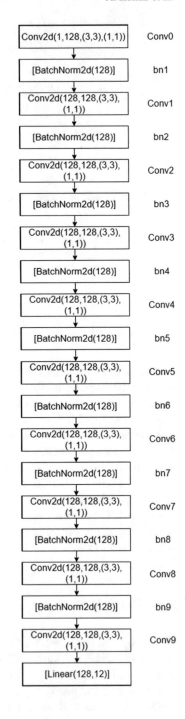

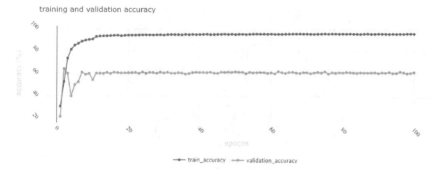

Fig. 16 Plot of training accuracy versus validation accuracy for ResNet_Mel

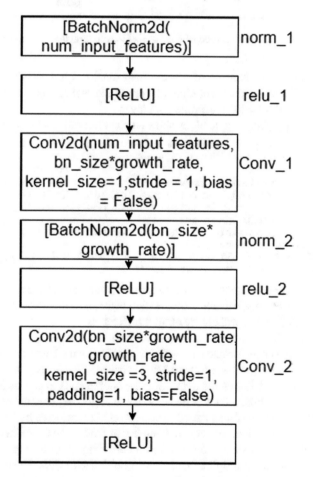

Fig. 17 Dense layer, a number of dense layers are stacked to form a dense block which is a part of DenseNet

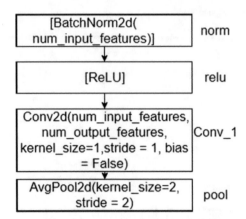

Fig. 18 Transition layer, which is used with the dense block to form the dense block

The transition layer is made of a two-dimensional batch normalization layer, a ReLU, a two-dimensional convolutional layer with kernel size = 1 and stride = 1, and finally a two-dimensional average pool layer.

The following PyTorch code block shows the configuration of DenseBlock.

```
for i, num_layers in enumerate(block_config):
    block = _DenseBlock(num_layers=num_layers,
num_input_features          =          num_features, bn_size =
bn_size, growth_rate = growth_rate, drop_rate = drop_rate)
    self.features.add_module('denseblock%d' % (i + 1), block)
    num_features = num_features + num_layers * growth_rate
    if i != len(block_config) - 1:
    trans = _Transition( num_input_features=num_features,
num_output_features=num_features // 2)
        self.features.add_module('transition%d' % (i + 1), trans)
        num_features = num_features // 2
```

Block_config consists of the number of layers of dense layers and transition layers in a DenseBlock. Finally, a batch normalization layer and a linear layer complete the architecture and give the probability of all the 12 labels.

The weights of convolutional layers are initiated using kaiming_normal distribution [39]; for the BatchNorm2d, the weights are filled with 1 and the bias is filled with 0. The output of first stack followed by the dense block gives us a tensor of size $(1024 \times 4 \times 4)$, which is then fed to the average_pool2d layer to give us the vector of size (1024); it is then finally fed to a linear classifier to get the predicted probability of the 12 labels. Number of filters added to each layer is represented by "growth_rate" ("k" in [3]). Num_int_features show the number of features to be learned in the first convolutional layer. Block_config is the number of layers in each pooling block, and bn_size shows the multiplicative factor of bottleneck layers. Using the MFCC feature extraction technique (Sect. 2.1) with the proposed DenseNet-like model, we achieved an accuracy of 97.58%. While using the mel feature extraction technique (Sect. 2.1) with the same model, we achieved an accuracy of 97.73%.

3.4 SeNet Models

The two major components of the SeNet-like model are the Pre-ActBlock (Fig. 19). The layer from the avgPool2d till last layer (sigmoid) in Fig. 19 forms the squeeze layer, as described in [5]. Then the output of the Pre-ActBlock (Fig. 19) (sigmoid layer) is multiplied with the output of conv2 (Fig. 19), and finally, the product thus

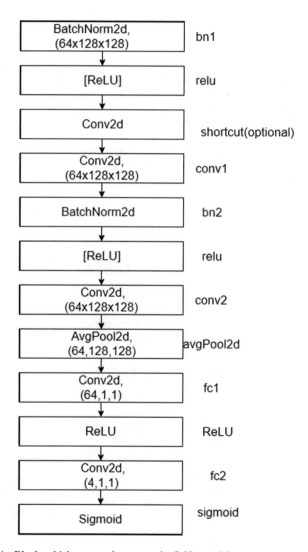

Fig. 19 Pre-ActBlock, which was used to create the SeNet models

obtained is added to the output of shortcut (optional) layer. The product followed by the sum completes the excitation layer as described in paper [5].

SeNet, whose block diagram is shown in Fig. 19, uses the feature extracted by mel (2.1) to achieve an accuracy of 97.45%, whereas using the MFCC feature extraction technique (3.1.1), we were able to achieve a score of 97.32%.

3.5   Ensemble

The performance of these models was also tested on speech recognition competition [40] public dataset. The accuracy and performance thus obtained are used to provide the weights to every model that we have developed. Taking the weighted average, we have ensemble all the models and thus achieved an accuracy of 98.098% (Fig. 20).

Fig. 20 SeNet model that is formed using the Pre-ActBlock shown in Fig. 19

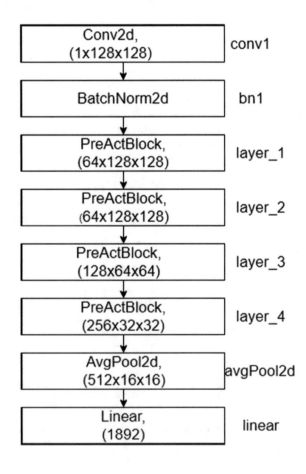

4 Experiments

4.1 Data

The dataset used for this project is described in [7]. It consists of 64,727, voice clips. Of which when developing the model, 6798 clips were used for validation and the rest 57,929 was used for training.

When training the model, finally we have retrained all the models with hyperparameters finalized during the development phase. We have used 57,982 voice clips for training, and then we have tested the individual models on 6835 voice clips.

The dataset was divided into 31 categories, out of which we used only 10 ("yes," "no," "up," "down," "left," "right," "on," "off," "stop," "go," "silence," "unknown") categories for testing our model, and we also added two categories: silence and unknown. Unknown is used to indicate the category which is not present in the 8 categories, leaving "Silence" aside.

4.2 Evaluation Metrics

We judged the performance of our model, based on the number of voice clips that it had correctly classified. Hence, simple measure of our model's performance is "the number of clips correctly classified"/"the total number of clips".

4.3 Experimental Setup

To carry out our experiment, we used the Kaggle instances [41]. It provides us with GPU required for training our model, 12 GB RAM, and all the required packages were preinstalled. We used PyTorch [42], the deep-learning framework for developing our solution.

All the models (10 in total) were trained for 100 epochs (some trained for 75 epochs). For the first 10 epochs, the learning rate used was 0.01, and after the 10 epochs, we used 0.001 as our learning rate. The loss function used for optimization was cross-entropy loss. We used the stochastic gradient descent (SGD) with the momentum of 0.9 and weight decay of 0.00001.

The training parameters are used for all the 10 models, and the accuracy obtained is shown in Table 1.

Table 1 Experiment configuration for different models

Model	Preprocessing	is_1d	Re-shape_size	Batch_size	Epochs	Accuracy
Vgg2d	Mel	False	128	32	100	97.76
Vgg2d	MFCC	False	128	32	100	97.60
Vgg1d	Raw	True	None	32	100	96.37
Vgg1d	Mel	True	None	64	100	97.22
ResNet	Mel	False	None	32	100	96.21
ResNet	MFCC	False	None	32	100	96.73
DenseNet121	Mel	False	128	16	100	97.73
DenseNet121	MFCC	False	128	16	100	97.58
SeNet	MFCC	False	128	16	75	97.32
SeNet	Mel	False	128	16	75	97.45

5 Analysis

Out of 6835 voice clips in test set, there are 48 voice clips that were misclassified by all the 10 models. On manual analysis, it was found that these voice clips are unrecognizable by the human ear as well.

There were certain voice clips that were recognized correctly by only one model, that too not by the model with the highest accuracy. Of example, the voice clip 1806.wav was classified by the model DenseNet with mel feature extraction technique. The correct label of 1806.wav was "go," which was not even recognized by human. Table 2 shows 15 such voice clips which were not recognizable human ears, but one of the 10 algorithms could recognize them correctly.

Table 3 shows the model size for different models.

Table 4 shows the execution time of all the models on test dataset.

Table 2 Sound clips that were not correctly recognized by humans but one of our models recognized them correctly

File name in dataset	Original label	Model with correct classification
five/94de6a6a_nohash_1.wav	Unknown (five)	SeNet with MFCC
go/3f2b358d_nohash_0.wav	go	vgg2d with MFCC
go/6205088b_nohash_1.wav	Go	DenseNet with Mel
left/3f2b358d_nohash_0.wav	left	vgg1d with raw
no/1cb788bc_nohash_1.wav	No	ResNet with mel
no/3f2b358d_nohash_0.wav	No	ResNet with mel
no/3f2b358d_nohash_1.wav	No	ResNet with mel
no/3f2b358d_nohash_2.wav	No	vgg2d with mel
right/3f2b358d_nohash_0.wav	Right	vgg1d_raw
right/3f2b358d_nohash_2.wav	Right	vgg1d_raw
right/5e3dde6b_nohash_1.wav	Right	vgg2d_mel
seven/3f2b358d_nohash_0.wav	Unknown	resnet_mel
up/9a7c1f83_nohash_0.wav	Up	vgg2d_mfcc
up/9a7c1f83_nohash_5.wav	Up	vgg2d_mfcc
yes/3f2b358d_nohash_0.wav	Yes	dense_net_mfcc

Table 3 Model size for different models

Model	Preprocessing	Model size (MB)
Vgg2d	Mel	41.2
Vgg2d	MFCC	41.2
Vgg1d	Raw	54.8
Vgg1d	Mel	39.0
ResNet	Mel	5.08
ResNet	MFCC	5.08
DenseNet121	Mel	27.0
DenseNet121	MFCC	27.0
SeNet	MFCC	43.3
SeNet	Mel	43.3

Table 4 Execution time of all the models

Model	Execution time (s)
Vgg2d	5.64
Vgg1d	0.32
Vgg1d	0.49
ResNet	0.19
DenseNet	1.55
SeNet	0.59

6 Conclusion

In this work, we started with voice clips and an approach to classify the keywords in using deep-learning techniques. We selected CNN to achieve our result, motivated by the paper [9]. It turned out that the architectures that had state-of-the-art performance for image recognition and also performed quite well on the small footprint keywords. As the keywords not have large dependencies, the CNN performed very well.

The hyperparameter space was explored only up to a limited extent because of the limited availability of GPU. The exploration of hyperparameter space (e.g., learning rate) can improve the results. To make the models adaptable for resource-constrained devices, we can try and limit the number of parameters to facilitate fast inference.

We have published the result as easy to train "Kaggle notebooks" [43] cloud instances.

References

1. Chen, G., Parada, C., Heigold, G.: Small-footprint keyword spotting using deep neural networks. In: Proceedings of ICASSP (2014)
2. https://link.springer.com/content/pdf/10.1186%2Fs13636-015-0068-3.pdf
3. Huang, G., Liu, Z., van der Maaten, L., Weinberger, K.Q.: DenseNet
4. He, K., Zhang, X., Ren, S., Sun, J.: Deep residual learning for image recognition. In: CVPR (2016)
5. Hu, J., Shen, L., Albanie, S., Sun, G., Wu, E.: https://arxiv.org/abs/1709.01507
6. Girshick, R., Donahue, J., Darrell, T., Malik, J.: Rich feature hierarchies for accurate object detection and semantic segmentation. In: CVPR (2014)
7. Warden, P.: Speech commands: a public dataset for single-word speech recognition
8. Schalkwyk, J., Beeferman, D., Beaufays, F., Byrne, B., Chelba, C., Cohen, M., Kamvar, M., Strope, B.: "Your word is my command": Google search by voice: a case study. In: Neustein, A. (ed.) Advances in Speech Recognition, pp. 61–90. Springer, USA (2010)
9. https://static.googleusercontent.com/media/research.google.com/en//pubs/archive/43969.pdf
10. LeCun, Y., Bengio, Y.: Convolutional networks for images, speech, and time-series. In: The Handbook of Brain Theory and Neural Networks. MIT Press (1995)
11. Abdel-Hamid, O., Mohamed, A., Jiang, H., Penn, G.: Applying convolutional neural network concepts to hybrid NN-HMM model for speech recognition. In: Proceedings of ICASSP (2012)
12. Toth, L.: Combining time- and frequency-domain convolution in convolutional neural network-based phone recognition. In: Proceedings of ICASSP (2014)
13. Sainath, T.N., Mohamed, A., Kingsbury, B., Ramabhadran, B.: Deep convolutional neural networks for LVCSR. In: Proceedings of ICASSP (2013)
14. LeCun, Y., Huang, F., Bottou, L.: Learning methods for generic object recognition with invariance to pose and lighting. In: Proceedings of CVPR (2004)
15. Russakovsky, O., Deng, J., Su, H., Krause, J., Satheesh, S., Ma, S., Huang, Z., Karpathy, A., Khosla, A., Bernstein, M., et al.: ImageNet large scale visual recognition challenge. IJCV
16. Perronnin, et al.: High-dimensional shallow feature encodings (the winner of ILSVRC-2011) (2010)
17. Zeiler, M.D., Fergus, R.: Visualizing and understanding convolutional networks. CoRR, abs/1311.2901 (2013). In: Proceedings of ECCV (2014)
18. Sermanet, P., Eigen, D., Zhang, X., Mathieu, M., Fergus, R., LeCun, Y.: OverFeat: integrated recognition, localization and detection using convolutional networks. In: Proceedings of ICLR (2014)

19. Howard, A.G.: Some improvements on deep convolutional neural network based image classification. In: Proceedings of ICLR (2014)
20. Szegedy, C., Liu, W., Jia, Y., Sermanet, P., Reed, S., Anguelov, D., Erhan, D., Vanhoucke, V., Rabinovich, A.: Going deeper with convolutions. In: CVPR (2015)
21. Ioffe, S., Szegedy, C.: Batch normalization: accelerating deep network training by reducing internal covariate shift. In: ICML (2015)
22. Simonyan, K., Zisserman, A.: Very deep convolutional networks for large-scale image recognition. In: ICLR (2015)
23. He, K., Zhang, X., Ren, S., Sun, J.: Delving deep into rectifiers: surpassing human-level performance on ImageNet classification. In: ICCV (2015)
24. He, K., Zhang, X., Ren, S., Sun, J.: Spatial pyramid pooling in deep convolutional networks for visual recognition. In: ECCV (2014)
25. Girshick, R.: Fast R-CNN. In: ICCV (2015)
26. Ren, S., He, K., Girshick, R., Sun, J.: Faster R-CNN: towards real-time object detection with region proposal networks. In: NIPS (2015)
27. Long, J., Shelhamer, E., Darrell, T.: Fully convolutional networks for semantic segmentation. In: CVPR (2015)
28. Bengio, Y., Simard, P., Frasconi, P.: Learning long-term dependencies with gradient descent is difficult. IEEE Trans. Neural Netw. **5**(2), 157–166 (1994)
29. Glorot, X., Bengio, Y.: Understanding the difficulty of training deep feedforward neural networks. In: AISTATS (2010)
30. LeCun, Y., Bottou, L., Orr, G.B., Müller, K.-R.: Efficient BackProp. In: Neural Networks: Tricks of the Trade, pp. 9–50. Springer (1998)
31. Saxe, A.M., McClelland, J.L., Ganguli, S.: Exact solutions to the nonlinear dynamics of learning in deep linear neural networks (2013). arXiv:1312.6120
32. LeCun, Y., Boser, B., Denker, J.S., Henderson, D., Howard, R.E., Hubbard, W., Jackel, L.D.: Backpropagation applied to handwritten zip code recognition. Neural Comput. **1**(4), 541–551 (1989)
33. Srivastava, R.K., Greff, K., Schmidhuber, J.: Training very deep networks. In: NIPS (2015)
34. He, K., Sun, J.: Convolutional neural networks at constrained time cost. In: CVPR (2015)
35. LeCun, Y., Bottou, L., Bengio, Y., Haffner, P.: Gradient-based learning applied to document recognition. Proc. IEEE **86**(11), 2278–2324 (1998)
36. https://pdfs.semanticscholar.org/c475/6dcc7afc2f09d61e6e4cf2199d9f6dd695cc.pdf?_ga=2.82886933.2025333462.1556684881-574832142.1549853665
37. https://www.kaggle.com/c/tensorflow-speech-recognition-challenge/discussion/47715#latest-294232
38. https://pdfs.semanticscholar.org/3de0/616eb3cd4554fdf9fd65c9c82f2605a17413.pdf
39. He, K., et al.: Delving deep into rectifiers: surpassing human-level performance on ImageNet classification (2015)
40. https://www.kaggle.com/c/tensorflow-speech-recognition-challenge/leaderboard
41. https://www.kaggle.com/kernels
42. Paszke, A., Gross, S., Chintala, S., Chanan, G., Yang, E., DeVito, Z., Lin, Z., Desmaison, A., Antiga, L., Lerer, A.: Automatic differentiation in PyTorch
43. https://www.kaggle.com/ashukr/models-for-keyword-spotting?scriptVersionId=14366175

Strategic Way to Count the Number of People in a Room Using Multiple Kinect Cameras

Yuchen Liu, Sourav Chakraborty, Ashutosh Kumar, Rakesh Seal
and Sohang Sengupta

Abstract This paper proposes a system in which we use multiple Kinect cameras placed in very strategic positions to accurately detect the presence of the number of human beings present in a room and also keep a count of the number entering and leaving the room. Banking physical security and enterprise data centre security are some of the areas where such technologies can effectively make a difference in terms of physical security practices and attendance.

Keywords Kinect camera · Face detection · OpenCV · Overhead camera · Deep learning · Facenet · Computer vision · Convolutional neural network (CNN)

Y. Liu
Department of Electrical and Computer Engineering, Boston University, Boston, USA
e-mail: lyc@bu.edu

S. Chakraborty
Imperial College London Business School, London, UK
e-mail: sourav.chakraborty18@imperial.ac.uk

A. Kumar
Department of Information Technology, Institute of Engineering and Management, Kolkata, India
e-mail: iashu12966@gmail.com

R. Seal (✉)
Department of Electronics and Communication Engineering, Institute of Engineering and Management, Kolkata, India
e-mail: rakeshseal0@gmail.com

S. Sengupta
Department of Computer Science and Engineering, Institute of Engineering and Management, Kolkata, India
e-mail: mailsohang@gmail.com

© Springer Nature Singapore Pte Ltd. 2020
M. Chakraborty et al. (eds.), *Proceedings of International Ethical Hacking Conference 2019*, Advances in Intelligent Systems and Computing 1065, https://doi.org/10.1007/978-981-15-0361-0_3

1   Introduction

There was a time not long ago that the biggest technological invention was the telephone, and the telephone then was wired and was used to allow people to communicate via speech. However as days progressed and technology became more advanced, there rose demands for even smarter technologies. Hence, the phone that we use today is more like a mini wireless computer that is used to perform a variety of functions besides just making voice calls. So what is the next thing that would employ the advances of technology and be considered in the SMART domain? Well, next-generation buildings are supposed to be equipped with SMART rooms. The rooms that would employ the operation of multiple sensors in order to adapt to automatically determine the heating, ventilation, lighting, audio–visual systems, etc., based on the number of people present in the room. Imagine a system that would automatically turn off the lights of the room when a person moves out of it and also control the HVAC to act accordingly, thereby tracking the movements of people. All the functionality of a SMART room is based solely on the ability to detect and track the presence of humans within a room using surveillance cameras. To this aim, we propose our project that employs two Kinect cameras to determine and track human heads and faces to keep a count of the occupancy within a small room and hence laying the first stepping-stone towards a better and a smarter tomorrow.

2   People Detection and Counting

The Kinect 1 is to be placed overhead the door. The Kinect 2 is placed inside the room but facing the door. When people move into the room, both the Kinects operate in collaboration. The task of the overhead Kinect 1 is to calculate the silhouette of the heads of the people moving in, while the face detection determines the number of face of the people moving in simultaneously. Both the cameras are used in a cross-check mechanism as the face detection camera won't be able to detect multiple faces of people if they are moving into the room in a line, as the faces of the people behind the first person will be covered by the former, whereas, on the contrary, the overhead Kinect 2 camera can determine the exact number of heads walking in. This collaborative effort would result in the correct estimation of the human occupancy within the room (Fig. 1).

2.1   Principle of the Overhead Camera

When capturing moving objects of a specific scene, the background image does not change within a period of time. Thus, by first subtracting the background image and extracting the moving objects, we can get the moving objects. After processing the

Fig. 1 Camera set-up

images, we could get the count of a number of people by the contour finding function of OpenCV. The background updating method is used to update the background in order to adapt to the environmental changes. The detection methods employed in this project are based on the following approaches.

Implementation of the overhead camera. Background subtraction: background subtraction (BS) is a common and widely used method to get the moving objects by using static cameras (Fig. 2).

Maintaining the integrity of the specifications. The implementation of the method is by firstly capturing a frame of the background image at time t. We use $I(t)$ to denote the image and use B to the background image. We are able to extract the objects by using subtraction. The pixel value of the current frame of image is denoted by $P[I(t)]$, and the pixel value of the background image is $P[B]$. By using $P[I(t)] - P[B]$, we can get the difference pixel value of a specific point. Therefore, by implementing this algorithm throughout the whole figure, we can extract the moving objects since same pixels (unchanged) have same value.

$$P[F(t)] = P[I(t)] - P[B]$$

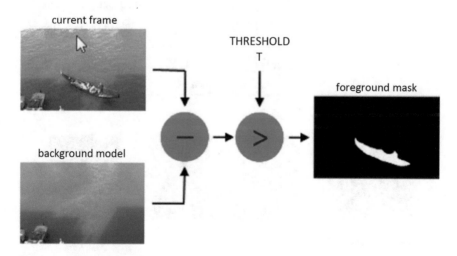

Fig. 2 Background subtraction

Since the background image is captured at time t, the calculation can only show the difference between the current frame of image and time t frame of image. Thus, when the background changes, it should be updated otherwise the calculation is wrong for some sense. We would use a threshold to show the difference of image, in this way, the result can be more clear to us.

$$|P[F(t)] - P[F(t + 1)]| > \text{Threshold}$$

When the subtraction large than threshold, which means the difference is obvious, then this pixel can be viewed as a changed one. Pixel difference larger than threshold will cause the pixel to be "white", otherwise "black". The threshold is dependent on movement speed. Faster objects means higher thresholds.

Smooth (Median filter). Median filter is a nonlinear filter in order to smooth the image and reduce noise. The median filter sets the grey value of each pixel to the median grey value of the surrounding pixels.

For median filtering, we use a specific window to move along the image. The median value of the pixels in the window is set as the output pixel value of these pixels. For example, if we have value of 2, 3, 4, 5, 6, we can get the value of current position to be 4 because it is the median value (Figs. 3, 4).

Dilation and then erosion can fill the hollow, link the nearby objects and smooth the boundaries. We use this method to process our image.

Rectangular to count the heads. We use cvFindcontours() and cvRect in the OpenCV library to find the desired objects.

Fig. 3 Median filter

123	125	126	130	140
122	124	126	127	135
118	120	150	125	134
119	115	119	123	133
111	116	110	120	130

Neighbourhood values:

115, 119, 120, 123, 124, 125, 126, 127, 150

Median value: 124

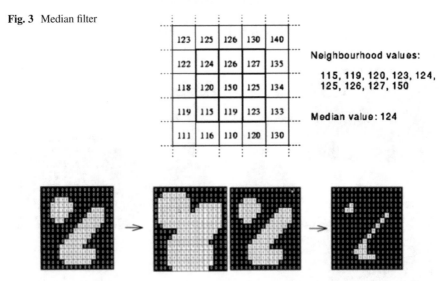

Fig. 4 Dilation and erosion

2.2 Front Camera (Face Detection)

Face detection can be regarded as a specific case of object-class detection. In object detection, the task is to find the locations and sizes of all objects in a digital image that belong to a given class. The front camera is set inside the room, facing towards the door. It counts the number of people by detecting human faces. The outcome of this camera along with the algorithm is to circle out the faces in the real-time video, also with the number of people detected shows in the backstage software.

Principle of the face detection. Here, we use the joint face detection using multitask cascaded convolutional networks [1] to detect number of faces captured by the front camera.

The cascade face detector proposed by Viola and Jones [2, 3] utilizes Haar-like features and AdaBoost to train cascaded classifiers, which achieve good performance with real-time efficiency. However, quite a few works [4, 5] indicate that this detector may degrade significantly in real-world applications with larger visual variations of human faces even with more advanced features and classifiers.

Recently, convolutional neural networks (CNNs) achieve remarkable progresses in a variety of computer vision tasks, such as image classification and face recognition. Inspired by the good performance of CNNs in computer vision tasks, some of the CNNs-based face detection approaches have been proposed in recent years.

The implemented CNNs consist of three stages. In the first stage, it produces candidate windows quickly through a shallow CNN. Then, it refines the windows to reject a large number of non-faces windows through a more complex CNN. Finally, it uses a more powerful CNN to refine the result and output facial landmarks positions. Thanks to this multitask learning framework, the performance of the algorithm can be notably improved.

Overall Framework. The overall pipeline of our approach is shown in Figs. 5, 6 and 7. Given an image, we initially resize it to different scales to build an image pyramid, which is the input of the following three-stage cascaded framework:

Stage 1: We exploit a fully convolutional network, called proposal network (P-Net), to obtain the candidate windows and their bounding box regression vectors. Then we use the estimated bounding box regression vectors to calibrate the candidates. After that, we employ non-maximum suppression (NMS) to merge highly overlapped candidates (Fig. 5).

Stage 2: All candidates are fed to another CNN, called refine network (R-Net), which further rejects a large number of false candidates, performs calibration with bounding box regression, and NMS candidate merge (Fig. 6).

Stage 3: This stage is similar to the second stage; but in this stage, we aim to describe the face in more details. In particular, the network will output five facial landmarks' positions (Fig. 7).

Implementation of the face detection. The implementation of this is pretty straightforward. The trained model was obtained from a popular open-source python library. The images were tested with this model in the tensor flow framework.

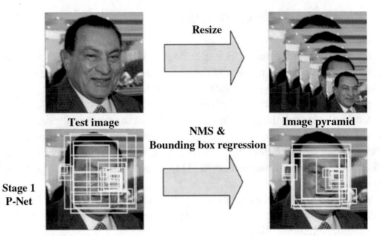

Fig. 5 P-net stage

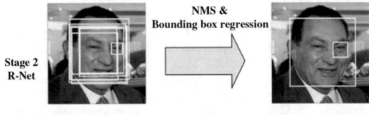

Fig. 6 R-net stage

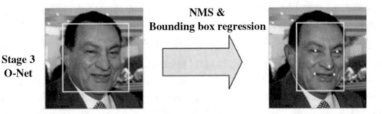

Fig. 7 O-net stage

After successfully detecting faces, the faces were marked with OpenCV for representation. Also, the number of faces was counted.

2.3 Experimental Results

Development environment. We use Win7 64 + Visual Studio 2013 + OpenCV 2.4.9 + Kinect 1 for overhead detection.

Manjaro linux 64bit + Tensorflow 1.13.1 + OpenCV 4.0.0.21 were used for face detection.

Overhead Detection Result. We can see from Figs. 8, 9, the edges on the human hair are smoothed in fig. 9. The median filter reduces the noise and smooth the image.

The final result is listed below, and the camera can easily detect one human figure, but when detecting two human figures, the bounding rectangles have some errors.

Face Detection Result. We have tested the program several times, according to the distance between camera and objects, camera angle, and lighting conditions, the test results deteriorated. Following is the experimental results (Figs. 10, 11).

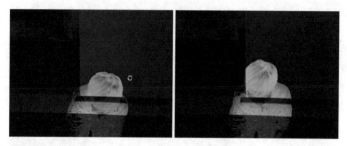

Fig. 8 Median filter

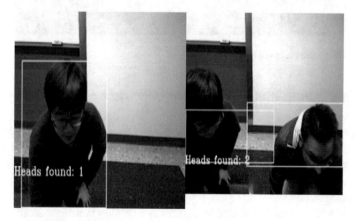

Fig. 9 Overhead detection result

Fig. 10 Three people result

Fig. 11 Multiple people result (more than 10)

2.4 Conclusions and Possible Improvements

The proposed approach mentioned above is the first step to achieving SMART rooms for the future. The SMART rooms would be an efficient office space for the generations to come and automation of the room environment based on the number of people present would ensure higher comfort, work throughput and efficient optimum power management.

The SMART rooms would also be a gateway to higher biometric security, where a present-day magnetic card would be replaced and multiple cameras detecting the human presence and recognizing individual faces to allow access and would ensure a higher security and tracking of human movements within an enclosed office space.

The facenet model can be used to detect faces from an existing database with a quite good accuracy. This can be used to control access in an office environment.

The face recognition can be used for automating the attendance system inside a classroom from an existing database of student images.

References

1. Zhang, K., Zhanpeng, Z., Li, Z., Qiao, Y.: Joint face detection and alignment using multi-task cascaded convolutional networks. IEEE Sign. Process. Lett. **23**(10), 1499–1503 (2016)
2. Viola, P., Jones, M.: Rapid Object Detection Using a Boosted Cascade of Simple Features. Cambridge, MA, USA (2001)
3. Viola, P., Jones, M.: Robust Real-Time Face Detection. Cambridge, MA, USA (2003)
4. Yang, B., Yan, J., Lei, Z., Li, S.Z.: Aggregate channel features for multi-view face detection. In: IEEE International Joint Conference on Biometrics, pp. 1–8 (2014)
5. Pham, M.T., Gao, Y., Hoang, V.D.D., Cham, T.J.: Fast polygonal integration and its application in extending haar-like features to improve object detection. In: IEEE Conference on Computer Vision and Pattern Recognition, pp. 942–949 (2010)
6. Xia, L., Chen, C.C., Aggarwal, J.K.: Human detection using depth information by Kinect. IEEE Comput. Vis. Pattern Recogn. Workshops, 2160–7508, 15–22, June (2001); Yamada, K., Mizuno, M.: A vehicle detection method using image segmentation. Electron. Commun. Japan Part 3. **84**(10) (2001)
7. Tian, Q., Zhou, B., Zhao, W., Wei, Y., Fei, W.: Human detection using hog features of head and shoulder based on depth map. J. Softw. **8**(9), 2223–2230; Yamada, K.: A vision sensor having an expanded dynamic range for autonomous vehicles. IEEE Trans. Veh. Technol. **47**, 332–341 (1998)
8. Zhang, L., Sturm, J., Cremers, D., Lee, D.: Real-time human motion tracking using multiple depth cameras. In: Proceedings of the International Conference on Intelligent Robot Systems (IROS) (2012)

Biomedical Image Security Using Matrix Manipulation and DNA Encryption

Mousomi Roy, Shouvik Chakraborty, Kalyani Mali, Arghasree Banerjee, Kushankur Ghosh and Sankhadeep Chatterjee

Abstract Biomedical image analysis is an integral part of the modern healthcare industry and has a huge impact on the modern world. Automated computer-aided systems are highly beneficial for fast, accurate and efficient diagnosis of the biomedical images. Remote healthcare systems allow doctors and patients to perform their jobs from separate geographic locations. Moreover, expert opinion about a patient can be obtained from a doctor who is in a different country or in some distant location within stipulated amount of time. Remote healthcare systems require digital biomedical images to be transferred over the network. But several security threats are associated with the transmission of the biomedical images. Privacy of the patients must be preserved by keeping the images safe from any unauthorized access. Moreover, the contents of the biomedical images must be preserved efficiently so that no one can tamper it. Data tampering can produce drastic results in many cases. In this work, a method for biomedical image security has been proposed. DNA encryption method is one of the emerging methods in the field of cryptography. A secure and

M. Roy · S. Chakraborty · K. Mali
Department of Computer Science & Engineering, University of Kalyani, Kalyani, Nadia, West Bengal, India
e-mail: iammouroy@gmail.com

S. Chakraborty
e-mail: shouvikchakraborty51@gmail.com

K. Mali
e-mail: kalyanimali1992@gmail.com

A. Banerjee · K. Ghosh · S. Chatterjee (✉)
Department of Computer Science & Engineering, University of Engineering & Management, Kolkata, West Bengal, India
e-mail: chatterjeesankhadeep.cu@gmail.com

A. Banerjee
e-mail: banerjeearghasree@gmail.com

K. Ghosh
e-mail: kush1999.kg@gmail.com

© Springer Nature Singapore Pte Ltd. 2020
M. Chakraborty et al. (eds.), *Proceedings of International Ethical Hacking Conference 2019*, Advances in Intelligent Systems and Computing 1065, https://doi.org/10.1007/978-981-15-0361-0_4

lossless encryption method is developed in this work. Various numerical parameters are used to evaluate the performance of the proposed method which proves the effectiveness of the algorithm.

Keywords Biomedical image analysis · DNA encryption · Remote health care · Cryptography

1 Introduction

To maintain the confidentiality of the data, it is important to use secured communication methods among several participating nodes. Modern world is highly dependent on the Internet [1], and there are several areas which cannot be imagined without Internet [2]. From banking to shopping, Internet provides several advantages in various ways. There, secured communication is a prime requirement of any online communication to make the data secured so that the confidentiality and data integrity of the users can be maintained [3]. Several types of data are used for communication purpose [4–8]. Image is one of the most important types of data that is used in several scenarios [9, 10]. Image transmission in defense, biomedical image communication [11–13] for remote health care [14, 15] and satellite image transmission are some of the frequently used examples, where images are the prime contents. Transmitting images without any protection can reveal important information and make the data vulnerable. Several methods are developed to provide security to the images. Various mathematical approaches are reported in the literature for image security [16–19]. DNA computing is one of the emerging research domains and has several applications in different fields, and among them, cryptography is a major one. DNA computing exploits the advantages of the biological complex phenomena and applies on the cryptography. It increases the security of the system compared to the conventional encryption methods that only relies on the complexity of the key and algorithm. In this article, biomedical images are taken into consideration. The security of the biomedical images is one of the major concerns in case of remote healthcare systems. Biomedical images demand high security because any kind of leakage can reveal sensitive information about patients [20]. Unauthorized person can misuse the sensitive biomedical information that can violate the confidentiality of the patient. Moreover, the integrity of the data is essential. Data can be tampered by the unauthorized persons if appropriate security measures are not imposed. In this work, the concept of DNA encryption is used with matrix manipulation to make the images secured. The remaining article is organized as follows. Section 2 illustrates the application and requirements of the DNA computing methods in encryption. Section 3 describes the proposed method in detail. Section 4 illustrates the obtained results and studies the performance of the proposed system. Section 5 concludes the work.

2 Application of the DNA Computing Concepts in Encryption

DNA computing is one of the most powerful methods that has several applications in various domains. Data security and cryptography are among them. DNA encryption exploits several advantages and concepts of biology to make the data more secure [21]. It mimics several biological phenomena in such a way so that the encryption becomes stronger than the conventional methods. In computer science, several problems are solved using bio-inspired methods and metaheuristic algorithms [22–32]. DNA encryption process can be successfully applied if it can model all the pixels of an image into the domain of DNA; i.e., all possible 256 combinations (in case of 8-bit grayscale images) must be mapped to its corresponding DNA sequence. Now, this mapping process should be well defined. Moreover, if the static mapping is used, then it can lead to some risks because anyone can trace the mapping pattern using cryptanalysis. Therefore, the mapping pattern must be changed in every transaction to make it safe against cryptographic attacks. The mapping can also be performed using a mapping matrix. The method must be robust to prevent various attacks. DNA encryption-based cryptographic methods must be designed in such a way so that digital environment can adopt the biological methods [33]. Many biological methods and concepts are difficult to mimic in digital environment and hence of no use. Therefore, during the design of the DNA-based encryption method, the feasibility or the implementation of the method is one of the prime concerns. DNA cryptography is dependent on the efficient modeling of the complex biological problems and exploits the difficulty to impose more security in the image. One of the most common attacks is the brute force attack. If one pattern always gets mapped to the same pattern for a certain mapping matrix, then it is vulnerable to brute force attacks. Moreover, modern cryptanalysis [34] methods may recover the actual pattern from it. Therefore, it is necessary to map multiple patterns of the same type to different cipher patterns. Moreover, it is also necessary for image encryption because if one plain pixel pattern gets mapped to a certain cipher pixel pattern then the correlation of the encrypted image will be high enough so that cryptanalysis methods can easily decipher it. Therefore, before designing any methods or framework for encryption, the abovementioned issues must be addressed; otherwise, the method can be susceptible to cryptographic attacks.

3 Proposed Method

A pixel of a grayscale image generally consists of 8 bits. Therefore, 256 combinations are possible for every pixel. Four characters, i.e., 'A', 'T', 'G' and 'C', are used in DNA computing. Therefore, to assign a unique DNA pattern for 256 possible combinations, at least four characters are needed for every combination (i.e., $4^4 = 256$). Now, any DNA pattern can be associated with any pixel pattern. In this way, a mapping matrix can be formed using which replacement can be performed. Now, this matrix can be manipulated in different ways to form the encoding matrix. The proposed algorithm is given below.

Algorithm 1: Proposed Algorithm

1. The sender decides the length of the DNA pattern 'l' (minimum is 4).
2. The sender prepares an encoding matrix.
3. Sender can perform any reversible operation on this matrix (e.g., transpose, inverse, etc.).
4. Sender receives an encoding matrix from the receiver.
5. Convert each pixel into binary.
6. For a particular pixel, find the corresponding mapping from both the matrices (i.e., the matrix generated by the sender and the matrix sent by the receiver).
7. Now to create the mapping of a pixel, take floor (l/2) from the DNA pattern of the first matrix and ceiling (l/2) from the DNA pattern of the second matrix and combine them.
8. Now, replace the sequence of every pixel with its complimentary sequence, i.e., 'A' with 'T', 'G' with 'C', 'T' with 'A' and 'C' with 'G'.
9. Convert the DNA sequence of each pixel into binary using the mapping table of the sender.
10. Perform XOR operation with each pair of pixels, and assign the value to the second pixel. A random binary sequence is generated for the first pixel to start with.
11. Convert the binary values into decimal to get the encrypted pixel.
12. Repeat steps 2 to 11 five times.

The proposed method is tested on different biomedical images using some standard quality measurement methods. The results are discussed in the next section.

4 Results and Discussion

The proposed algorithm is used to encrypt some biomedical images, and the results of the encryption are evaluated using some standard quality evaluation parameters [35]. Five different types of images are chosen to test the proposed method. These types are X-ray [36], CT scan [37], MRI [38], USG [39] and histological image [40]. Figures 1, 2, 3, 4 and 5 show the results of the encryption.

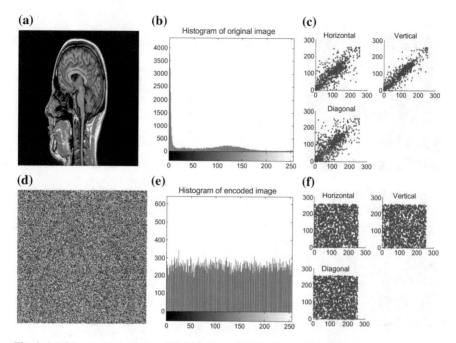

Fig. 1 MRI image encryption. **a** Original image. **b** Histogram of the original image. **c** Correlations of the original image (horizontal, vertical, diagonal). **d** Encrypted image. **e** Histogram of the encrypted image. **f** Correlations of the encrypted image (horizontal, vertical, diagonal)

4.1 Correlation Coefficients

Correlation coefficient is evaluated for three different directions, i.e., horizontal, vertical and diagonal. Correlation coefficients are calculated for the selected pairs using Eq. 1.

$$R_{xy} = \text{COV}(xy)/\sqrt{D(x)}\sqrt{D(y)} \tag{1}$$

where

$$\text{COV}(xy) = \frac{1}{T}\sum_{i=1}^{T}((x_i - E(x))(y_i - E(y))) \tag{2}$$

$$E(x) = \frac{1}{T}\sum_{i=1}^{T}x_i, \quad E(y) = \frac{1}{T}\sum_{j=1}^{T}y_j \tag{3}$$

$$D(x) = \frac{1}{T}\sum_{i=1}^{T}(x_i - E(x_i))^2, \quad D(y) = \frac{1}{T}\sum_{i=1}^{T}(y_i - E(y_i))^2 \tag{4}$$

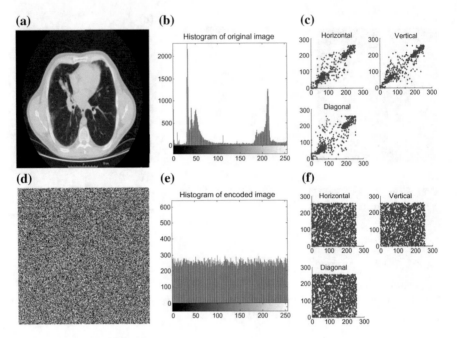

Fig. 2 CT scan image encryption. **a** Original image. **b** Histogram of the original image. **c** Correlations of the original image (horizontal, vertical, diagonal). **d** Encrypted image. **e** Histogram of the encrypted image. **f** Correlations of the encrypted image (horizontal, vertical, diagonal)

where x and y denote the grayscale values of the two adjacent pixels, and T is the total pair of pixels randomly selected from the image. Table 1 shows the values of correlation coefficients on different images.

4.2 PSNR

PSNR stands for peak signal-to-noise ratio. PSNR is a well-known parameter and can be computed from Eq. 5.

$$\text{PSNR} = 10 \log_{10}\left(\frac{L^2}{\text{MSE}}\right) \tag{5}$$

where

$$\text{MSE} = \frac{1}{N} \sum_{i=0,\, j=0}^{N,N} \left(x_{ij} - y_{ij}\right)^2 \tag{6}$$

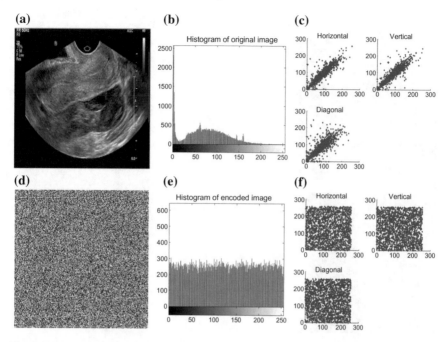

Fig. 3 USG image encryption. **a** Original image. **b** Histogram of the original image. **c** Correlations of the original image (horizontal, vertical, diagonal). **d** Encrypted image. **e** Histogram of the encrypted image. **f** Correlations of the encrypted image (horizontal, vertical, diagonal)

N is the pixel count, and x_{ij}, y_{ij} are the ith and jth pixels in the original and processed image. L ranges from 0 to 255 for grayscale images. Table 2 shows the PSNR values for different images.

4.3 Differential Attack

To assess the effect of the small change in original image on the encrypted image, two well-known parameters are used: number of pixels change rate (NPCR) and unified average changing intensity (UACI) which are defined in Eqs. 7 and 8, respectively.

$$\text{NPCR} = \frac{\sum_{i=j=1}^{m,n} D(i,j)}{w \times h} \times 100\% \qquad (7)$$

$$\text{UACI} = \frac{1}{w \times h} \left[\sum_{i,j}^{m,n} \frac{|C_1(i,j) - C_2(i,j)|}{255} \right] \times 100\% \qquad (8)$$

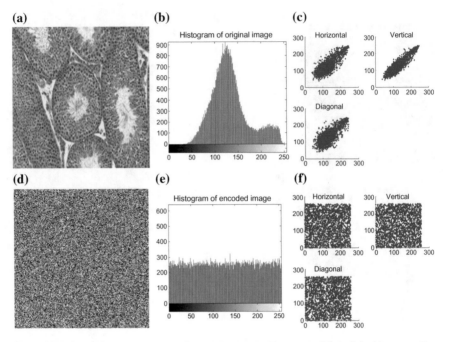

Fig. 4 Histogram image encryption. **a** Original image. **b** Histogram of the original image. **c** Correlations of the original image (horizontal, vertical, diagonal). **d** Encrypted image. **e** Histogram of the encrypted image. **f** Correlations of the encrypted image (horizontal, vertical, diagonal)

Here, C_1 and C_2 are two images after encryption with one pixel difference, w and h denote the image width and height, and $D(i, j)$ can be computed using on Eq. 9.

$$D(i, j) = \begin{Bmatrix} 1 \text{ if } C_1(i, j) = C_2(i, j) \\ 0 \text{ Otherwise} \end{Bmatrix} \tag{9}$$

Table 3 shows the NPCR and UACI values for different images.

5   Conclusion

Biomedical image security is one of the prime concerns in today's world. DNA-based computing has a huge prospect in the field of cryptography. In this work, a DNA computing-based new method is proposed to encrypt biomedical images. The proposed method is simple enough to be implemented. The analysis of this method proves that the proposed method can be used in real-world applications. There are several domains in which this method can be extended. Medical IoT (MIoT)-based systems are one of the most important applications of the proposed system where the proposed system can be proven to be efficient enough to keep the privacy of the

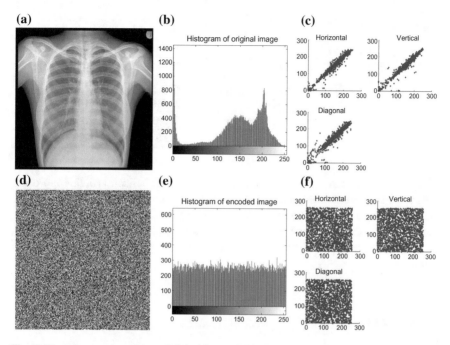

Fig. 5 X-ray image encryption. **a** Original image. **b** Histogram of the original image. **c** Correlations of the original image (horizontal, vertical, diagonal). **d** Encrypted image. **e** Histogram of the encrypted image. **f** Correlations of the encrypted image (horizontal, vertical, diagonal)

Table 1 Correlation coefficients in original and encrypted images after applying the proposed encryption algorithm

Test image	Horizontal		Vertical		Diagonal	
	Original	Encrypted	Original	Encrypted	Original	Encrypted
MRI	0.9299	0.0003	0.9647	0.0030	0.9075	−0.0047
CT scan	0.9795	−0.0006	0.9784	0.0011	0.9614	−0.0121
USG	0.9425	0.0155	0.9444	−0.0021	0.8916	−0.0049
Histopathology	0.8287	0.0092	0.9031	−0.0026	0.7763	0.0053
X-ray	0.9929	−0.0052	0.9912	−0.0012	0.9849	−0.0031

Table 2 PSNR values obtained after applying the proposed method

Test image	PSNR	
	Original–decrypted	Original–encrypted
MRI	Undefined	5.9377
CT scan	Undefined	6.6200
USG	Undefined	1.3379
Histopathology	Undefined	9.5984
X-ray	Undefined	7.6355

Table 3 NPCR and UACI values of some standard images obtained using the proposed approach

Test image	Proposed approach	
	NPCR (%)	UACI (%)
MRI	99.6231	42.1072
CT scan	99.6200	38.3748
USG	99.5101	37.2157
Histopathology	99.6520	27.6672
X-ray	99.6017	33.8719

patients. Sensitive information can be protected with the application of this approach. This work can be extended in terms of the algorithm or the application domain. The proposed method can be modified and can be hybridized with other methods. For example, the random number can be generated using the theory of chaos. Hence, there is an opportunity to extend this work and can be applied on different types of data.

References

1. Sarddar, D., Chakraborty, S., Roy, M.: An efficient approach to calculate dynamic time quantum in round Robin algorithm for efficient load balancing. Int. J. Comput. Appl. **123**, 48–52 (2015). https://doi.org/10.5120/ijca2015905701
2. Datta, S., Chakraborty, S., Mali, K., Baneijee, S., Roy, K., Chatterjee, S., Chakraborty, M., Bhattacharjee, S.: Optimal usage of pessimistic association rules in cost effective decision making. In: 2017 4th International Conference on Opto-Electronics and Applied Optics. Optronix 2017 (2018). https://doi.org/10.1109/optronix.2017.8349976
3. Wahballa, O., Wahaballa, A., Li, F., Idris, I.: C.X.-I.N. security, undefined 2017, medical image encryption scheme based on Arnold transformation and ID-AK protocol. Int. J. Netw. Secur. (n.d.)
4. Hore, S., Chakraborty, S., Chatterjee, S., Dey, N., Ashour, A.S., Van Chung, L., Le, D.-N.: An integrated interactive technique for image segmentation using stack based seeded region growing and thresholding. Int. J. Electr. Comput. Eng. **6** (2016). https://doi.org/10.11591/ijece.v6i6.11801
5. Hore, S., Chakroborty, S., Ashour, A.S., Dey, N., Ashour, A.S., Sifaki-Pistolla, D., Bhattacharya, T., Chaudhuri, S.R.B.: Finding contours of hippocampus brain cell using microscopic image analysis. J. Adv. Microsc. Res. **10**, 93–103 (2015). https://doi.org/10.1166/jamr.2015.1245
6. Chakraborty, S., Roy, M., Hore, S.: A study on different edge detection techniques in digital image processing (2018). https://doi.org/10.4018/978-1-5225-5204-8.ch070
7. Chakraborty, S., Roy, M., Hore, S.: A study on different edge detection techniques in digital image processing (2016). https://doi.org/10.4018/978-1-5225-1025-3.ch005
8. Roy, M., Chakraborty, S., Mali, K., Chatterjee, S., Banerjee, S., Mitra, S., Naskar, R., Bhattacharjee, A.: Cellular image processing using morphological analysis. In: 2017 IEEE 8th Annual Ubiquitous Computing Electronics and Mobile Communication Conference. UEMCON 2017 (2018). https://doi.org/10.1109/uemcon.2017.8249037
9. Chakraborty, S., Chatterjee, S., Dey, N., Ashour, A.S., Ashour, A.S., Shi, F., Mali, K.: Modified cuckoo search algorithm in microscopic image segmentation of hippocampus. Microsc. Res. Tech. **80** (2017). https://doi.org/10.1002/jemt.22900

10. Hore, S., Chatterjee, S., Chakraborty, S., Shaw, R.K.: Analysis of Different Feature Description Algorithm in object Recognition, pp. 66–99 (n.d.). https://doi.org/10.4018/978-1-5225-1025-3.ch004
11. Chakraborty, S., Chatterjee, S., Ashour, A.S., Mali, K., Dey, N.: Intelligent computing in medical imaging: a study. In: Dey, N. (ed.) Advancements in Applied Metaheuristic Computing. IGI Global, pp. 143–163 (2017). https://doi.org/10.4018/978-1-5225-4151-6.ch006
12. Chakraborty, S., Chatterjee, S., Dey, N., Ashour, A.S., Shi, F.: Gradient approximation in retinal blood vessel segmentation. In: 2017 4th IEEE Uttar Pradesh Section International Conference on Electrical, Computer and Electronics. IEEE, pp. 618–623 (2017). https://doi.org/10.1109/upcon.2017.8251120
13. Chakraborty, S., Mali, K., Chatterjee, S., Banerjee, S., Roy, K., Dutta, N., Bhaumik, N., Mazumdar, S.: Dermatological effect of UV rays owing to ozone layer depletion. In: 2017 4th International Conference on Opto-Electronics and Applied Optics. Optronix 2017 (2018). https://doi.org/10.1109/optronix.2017.8349975
14. Chakraborty, S., Mali, K., Banerjee, S., Roy, K., Saha, D., Chatterjee, S., Dutta, S., Majumder, S.: Bag-of-features based classification of dermoscopic images. In: 2017 4th International Conference on Opto-Electronics and Applied Optics. IEEE, pp. 1–6 (2017). https://doi.org/10.1109/optronix.2017.8349977
15. Chakraborty, S., Mali, K., Chatterjee, S., Banerjee, S., Roy, K., Deb, K., Sarkar, S., Prasad, N.: An integrated method for automated biomedical image segmentation. In: 2017 4th International Conference on Opto-Electronics and Applied Optics. IEEE, pp. 1–5 (2017). https://doi.org/10.1109/optronix.2017.8349978
16. Chakraborty, S., Seal, A., Roy, M., Mali, K.: A novel lossless image encryption method using DNA substitution and chaotic logistic map. Int. J. Secur. Appl. 10 (2016). https://doi.org/10.14257/ijsia.2016.10.2.19
17. Seal, A., Chakraborty, S., Mali, K.: A new and resilient image encryption technique based on pixel manipulation, value transformation and visual transformation utilizing single–level haar wavelet transform (2017). https://doi.org/10.1007/978-981-10-2035-3_61
18. Mali, K., Chakraborty, S., Seal, A., Roy, M.: An efficient image cryptographic algorithm based on frequency domain using Haar wavelet transform. Int. J. Secur. Appl. 9, 279–288 (2015). https://doi.org/10.14257/ijsia.2015.9.12.26
19. Roy, M., Mali, K., Chatterjee, S., Chakraborty, S., Debnath, R., Sen, S.: A study on the applications of the biomedical image encryption methods for secured computer aided diagnostics. In: 2019 Amity International Conference on Artificial Intelligence, IEEE, pp. 881–886 (2019). https://doi.org/10.1109/aicai.2019.8701382
20. Zhang, S., Gao, T., Gao, L.: A novel encryption frame for medical image with watermark based on hyperchaotic system. Math. Probl. Eng. 2014, 1–11 (2014). https://doi.org/10.1155/2014/240749
21. Sreeja, C.S., Misbahuddin, M., Mohammed Hashim, N.P.: DNA for information security: a survey on DNA computing and a pseudo DNA method based on central dogma of molecular biology. Int. Conf. Comput. Commun. Technol. IEEE, 1–6 (2014). https://doi.org/10.1109/iccct2.2014.7066757
22. Chakraborty, S., Bhowmik, S.: Blending roulette wheel selection with simulated annealing for job shop scheduling problem. In: Michael Faraday IET International Summit 2015, Institution of Engineering and Technology, p. 100(7) (2015) https://doi.org/10.1049/cp.2015.1696
23. Chakraborty, S., Bhowmik, S.: An efficient approach to job shop scheduling problem using simulated annealing. Int. J. Hybrid Inf. Technol. 8, 273–284 (2015). https://doi.org/10.14257/ijhit.2015.8.11.23
24. Chakraborty, S., Bhowmik, S.: Job shop scheduling using simulated annealing. In: First International Conference on Computation and Communication Advancement, pp. 69–73. McGraw Hill Publication (2013). https://scholar.google.co.in/citations?user=8lhQFaYAAAAJ&hl=en. Accessed 24 Nov 2017
25. Chakraborty, S., Seal, A., Roy, M.: An elitist model for obtaining alignment of multiple sequences using genetic algorithm. In: 2nd National Conference NCETAS 2015. Int. J. Innov. Res. Sci. Eng. Technol. 61–67 (2015)

26. Chakraborty, S., Mali, K., Chatterjee, S., Banerjee, S., Mazumdar, K.G., Debnath, M., Basu, P., Bose, S., Roy, K.: Detection of skin disease using metaheuristic supported artificial neural networks. In: 2017 8th Industrial Automation and Electromechanical Engineering. Conference. IEMECON 2017 (2017). https://doi.org/10.1109/iemecon.2017.8079594

27. Chakraborty, S., Mali, K., Chatterjee, S., Anand, S., Basu, A., Banerjee, S., Das, M., Bhattacharya, A.: Image based skin disease detection using hybrid neural network coupled bag-of-features. In: 2017 IEEE 8th Annual Ubiquitous Computing, Electronics and Mobile Communication Conference. UEMCON 2017 (2018). https://doi.org/10.1109/uemcon.2017.8249038

28. Roy, M., Chakraborty, S., Mali, K., Chatterjee, S., Banerjee, S., Chakraborty, A., Biswas, R., Karmakar, J., Roy, K.: Biomedical image enhancement based on modified Cuckoo search and morphology. In: 2017 8th Industrial Automation and Electromechanical Engineering Conference. IEMECON 2017 (2017). https://doi.org/10.1109/iemecon.2017.8079595

29. Chakraborty, S., Mali, K., Chatterjee, S., Banerjee, S., Sah, A., Pathak, S., Nath, S., Roy, D.: Bio-medical image enhancement using hybrid metaheuristic coupled soft computing tools. In: 2017 IEEE 8th Annual Ubiquitous Computing, Electronics and Mobile Communication Conference. UEMCON 2017 (2018). https://doi.org/10.1109/uemcon.2017.8249036

30. Chakraborty, S., Mali, K.: Application of multiobjective optimization techniques in biomedical image segmentation—a study. In: Multi-objective Optimization. Springer, Singapore, pp. 181–194. https://doi.org/10.1007/978-981-13-1471-1_8

31. Chakraborty, S., Raman, A., Sen, S., Mali, K., Chatterjee, S., Hachimi, H.: Contrast optimization using elitist metaheuristic optimization and gradient approximation for biomedical image enhancement. In: 2019 Amity International Conference on Artificial Intelligence. IEEE, pp. 712–717 (2019). https://doi.org/10.1109/aicai.2019.8701367

32. Chakraborty, S., Chatterjee, S., Chatterjee, A., Mali, K., Goswami, S., Sen, S.: Automated breast cancer identification by analyzing histology slides using metaheuristic supported supervised classification coupled with bag-of-features. In: 2018 Fourth International Conference on Research in Computational Intelligence and Communication Networks, IEEE, pp. 81–86 (2018). https://doi.org/10.1109/icrcicn.2018.8718736

33. Kaundal, A.K., Verma, A.K.: DNA Based Cryptography: A Review (2014). http://www.irphouse.com. Accessed 11 June 2019

34. Li, S., Zheng, X.: Cryptanalysis of a chaotic image encryption method. In: Proceedings of the IEEE International Symposium on Circuits System, vol. 2, pp. 708–711 (2002). https://doi.org/10.1109/iscas.2002.1011451

35. Mali, K., Chakraborty, S., Roy, M.: A study on statistical analysis and security evaluation parameters in image encryption. IJSRD Int. J. Sci. Res. Dev. 3, 2321–2613 (2015). www.ijsrd.com. Accessed 16 July 2018

36. Digital X-ray—Sand Lake Imaging—Radiology Centers: (n.d.). https://www.sandlakeimaging.com/procedures/digital-x-ray/. Accessed 14 July 2019

37. CT Scan for the Chest—Cedars-Sinai: (n.d.). https://www.cedars-sinai.edu/Patients/Programs-and-Services/Imaging-Center/For-Patients/Exams-by-Procedure/CT-Scans/CT-Chest.aspx. Accessed 14 July 2019

38. MRI Scan in Fayetteville, NC | Valley Radiology: (n.d.). https://www.valleyradiologync.com/services/mri-scan. Accessed 14 July 2019

39. Lee, R., Dupuis, C., Chen, B., Smith, A., Kim, Y.H.: Diagnosing ectopic pregnancy in the emergency setting. Ultrasonography 37, 78–87 (2018). https://doi.org/10.14366/usg.17044

40. Anatomy and Neuroscience: (n.d.). https://biomedicalsciences.unimelb.edu.au/departments/anatomy-and-neuroscience. Accessed 14 July 2019

Blockchain and AI

A Study on the Issue of Blockchain's Energy Consumption

Eshani Ghosh and Baisakhi Das

Abstract Blockchain technology is one of the biggest innovative technology that has been developed and has potential usage in fields of education, business and industries. Since the creation of bitcoins, blockchain has emerged as a means for storing digital information without the intervention of any third parties. However, now, it is used for various other applications than just being a simple distributed ledger. With time, it has imposed a larger impact on different fields of economy and has gained popularity for its immutability. But, there are some issues faced by blockchain. One of such issues is the energy consumption. Blockchains are found to consume exorbitant amount of energy because of the algorithm followed for its creation. This paper explores the blockchain technology and the impacts of energy consumption due to the technology used.

Keywords Blockchain · Proof of work · Energy consumption

1 Introduction

Being an immutable distributed ledger, blockchains have gained popularity in the recent years. It was first introduced as a cryptocurrency (bitcoin) by Satoshi Nakamoto [1]. He found bitcoin to be a digital way of transacting currencies using the peer-to-peer network policy and presenting bitcoin as a highly secure system for strong transaction histories without the intervention of any third parties [2]. The technology behind the development of the bitcoin, known as blockchain, became popular and emerged as a powerful system for storing any kind of digital information [1].

The most attractive features of the blockchain are its irresistibility to any information leakage. This property is acquired from cryptographic algorithms which are

E. Ghosh (✉) · B. Das
Department of Information Technology, Institute of Engineering and Management,
Salt Lake Electronics Complex, Sector V, Kolkata, India
e-mail: ghosh.eshani34@gmail.com

B. Das
e-mail: baisakhi83@gmail.com

© Springer Nature Singapore Pte Ltd. 2020
M. Chakraborty et al. (eds.), *Proceedings of International Ethical Hacking Conference 2019*, Advances in Intelligent Systems and Computing 1065, https://doi.org/10.1007/978-981-15-0361-0_5

63

used to join the blocks in the chain. Also, its peer-to-peer network communication and decentralised structure have strengthened the security. Blockchain architecture uses proof-of-work mechanism to mine a new block and stores the digital information permanently in the blockchain. The architecture of blockchain has paved the way for many applications [3]. Companies are nowadays trying to use blockchains for storing employee records, project records, etc. It is also used in the medical field for storing various medical records. Its use is not only limited to this, in broader sense, it can be used for implementing IoTs, storing legal information, supply chain management, etc., in the near future [4]. Altogether, it can be used for digitalizing the world in a protected manner.

Blockchain technology, being so advantageous, faces many challenges in its implementation. One of such challenges is the energy consumption issue. It has been found that blockchains, for enhancing its security, is consuming enormous amount of energy, and the root cause behind this is the proof-of-work mechanism. The proof-of-work, a mechanism to mine block, is found to consume maximum amount of energy out of the entire blockchain architecture. This is of great concern as the sources of energy are the non-renewable resources, so solving the energy consumption issue is the need of the hour for implementing blockchains. This paper discusses a survey on the blockchain architecture, the issues on energy consumption and methods to overcome the exorbitant energy consumption. Section 2 explains the blockchain architecture. The issues on energy consumption of blockchain are defined in Sect. 3. Section 4 explains the methods to reduce energy consumption, and Sect. 5 concludes the paper.

2 Blockchain Architecture

Blockchain architecture uses the mechanism of distributing the digital information rather than copying it. It is intended to timestamp digital documents so that it is not possible to tamper with them, whenever information is recorded. Blockchains can be defined as a chain of blocks which can hold certain records, and the blocks are linked to one another using the principles of cryptography. Each block consists of three main sections—data, hash and hash of previous block. The data that is stored depends on the type of blockchain used. For instance, if it is a cryptocurrency, then it stores information about the transactions that have occurred. Hash in a block is the fingerprint and is unique to that particular block. If any changes are made to the block, then the hash value also changes. So, hashes can be used to detect changes to a block. Hash of the previous block helps to create the chain and makes the system more secure as shown in Fig. 1. Block 2 contains its own hash function (2BIF) and hash of block 1 (AQCN) as the previous hash and so on and hence creates the chain of blocks. Whenever someone changes the hash value of the block, then that hash value needs to be updated in the next block. The next block will have a changed hash value due to the insertion of a new previous hash value. This is how the entire chain has to be updated to change a particular block which is practically impossible and hence

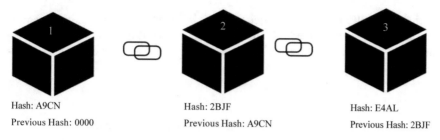

Hash: A9CN

Previous Hash: 0000

Hash: 2BJF

Previous Hash: A9CN

Hash: E4AL

Previous Hash: 2BJF

Fig. 1 Structure of Blockchain [2]

improves the security of the system. That is why blockchains are more commonly known as immutable ledger because it is not possible to tamper with them.

2.1 Blockchains—An Immutable Ledger

As stated earlier, blockchains are distributed, decentralised and immutable ledger system. In blockchain, the blocks are connected to each other via links which are established by the hash of each block.

If any change is made to the block, then the entire blockchain needs to be updated. So, blockchains are said to be immutable ledger. Figure 2 explains how a digital data is certified using blockchain. The data is kept inside a blockchain, and it generates a

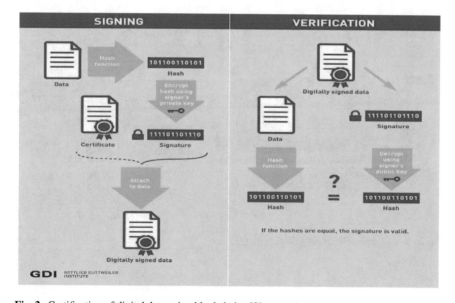

Fig. 2 Certification of digital data using blockchains [2]

hash value for that newly added block. The data is signed by encrypting the hash value and then stamped as certified [2]. Then, the verification process is done by decrypting the hash value and matching it with the one that is generated. If it matches, it means the data got verified and is authentic.

2.2 Blockchain as a Distributed Peer-to-Peer Network

There can be a situation where someone tampers with a block and also updates the entire blockchain to make the changes acceptable for all the blocks. This can result in information hacking or changing or even deletion of a block. To prevent such tampering, blockchain architecture has a distributed peer-to-peer network system. The blockchain is not only available in one computer, but it is distributed to all the computers in a network. So, any change made to the blockchain that information gets communicated to all the computers in a network. So, if a hacker changes the contents of a block and also modifies the immediately following blocks' previous hash value, then that modification is compared with blockchains in other computers in a network. If any change is detected in one blockchain, then that is cancelled and the previous state of blockchain is preserved. This helps in making the system more secure. Figure 3 shows how the peer-to-peer network is working when a transaction takes place and a new block gets added to the blockchain [5]. The block is then shared with all other computers in a network.

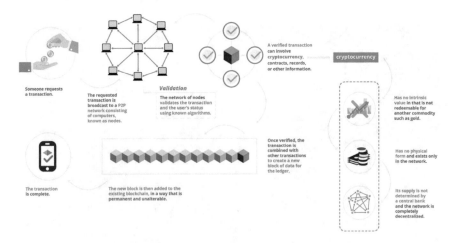

Fig. 3 Simple transaction using blockchain [5]

2.3 Proof-of-Work

It is a protocol which works by asking all the nodes in the network to solve a cryptographic puzzle by using the brute force algorithm [3]. Whenever a block is to be added, it should have a hash.

The hash is a 64-bit hexadecimal value which is generated using the sha256 algorithm. There is a field called number used only once (nonce) which is present in every block along with the hash, previous hash and the block number as shown in Fig. 4. There is a range generated within which the hash value of the block should lie. On changing the nonce value, the hash value that is generated is checked if it matches the required condition for block generation or not. This process of checking consumes energy along with that the values of hash that is generated require long sequence of arithmetic operations that need a lot of energy to be evaluated. Also, the values of hash abruptly change with the change in nonce, and so, a lot of evaluations are involved in this step. Whenever a miner solves the cryptographic puzzle (the entire process of hash generation), a new block is added to the blockchain and the block is transmitted and added to all the blockchains in the different nodes using peer-to-peer communication and this is how all blockchains get the new block added to them. But the generation of each block involves a lot of arithmetic computations which consumes a lot of energy. So, in blockchains, the mining process is the root cause for the energy consumption issue behind it.

3 Issues on the Energy Consumption of Blockchains

One of the major issues faced by blockchains is that it consumes exorbitant amount of energy mainly during the process of mining. The algorithm used for evaluating proof-of-work needs to be executed multiple times to match the target value and as the hash value changes in a non-uniform way, so it is completely based on trial and error method. For instance, in the case of bitcoin, miners take about ten minutes to mine a new block. This process has surely induced security but at the cost of enormous energy consumption.

Fig. 4 Structure of a block

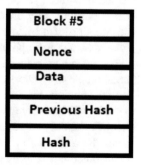

In an article by Steven Huckle in 2016, named Socialism and Blockchain, we get an alarming information. In this article, we get an estimate that bitcoin mining's annual energy consumption is 3.38 Terawatt Hours (TWH). This enormous amount of energy is equal to the total annual consumption of Jamaica in the year 2014 [6]. As stated by Hern [7], the energy consumption of the entire bitcoin network is found to be higher than Ireland. Study says that bitcoin will use 0.5% of electricity in the world by the end of 2018, as we know that the electricity demand comes from complex computing and with days more complex puzzles need to be solved. So, this has become a threat to the environment for its enormous amount of energy consumption which in turn increases the carbon footprint [7].

The mining cost of metals like gold is very high because of its extraction and its demand in the markets. Similarly, energy cost associated with mining of blocks is high because of the proof-of-work mechanism used. According to a study by Oak Ridge Institute in Cincinnati, it has been found that energy cost of mining of bitcoins is nearly 7 megajoules of energy which is equivalent to mining platinum [8]. Out of all the cryptocurrencies, bitcoin consumes the maximum energy as compared with the energy equivalent to mining copper, platinum and gold and the energy cost for further mining increases over time [8]. Figure 7 shows that the cryptocurrencies mining needs more energy per dollar generated compared to mining of copper, platinum and gold.

However, the annual consumption of this currency is rising in an exponential manner, currently which has approached a colossal amount of 55 TWH. This is obviously a matter of concern. Huckle [6] discusses that between 3 and 13 m metric tonnes of CO_2 were released in the environment due to bitcoin mining, that too only in the first half of 2018. If the utilization of bitcoin keeps on increasing, then severe consequences might be faced by the future generations. The effect of emission is such that it can contribute in an overall global temperature rise of 2 °C, as approximated by a team in Hawaii. Not only this, the global money supply in circulation is estimated at $11,000 billion. This will lead to corresponding energy consumption exceeding a huge amount 4000 GW [9]. This insane amount of energy is eight times the electricity consumption of France and twice the USA. The bitcoin can, therefore, be the burden for the climate.

Figure 5 shows the relative energy consumption of bitcoins in different countries where the X-axis is representing the percentage that could be powered by bitcoin, and the Y-axis is showing the various countries. According to this, Czech Republic has the maximum percentage of energy consumption, whereas USA has the least energy consumed by bitcoins. A statistic of 2018 [2] in Table 1 shows the energy consumption of a bitcoin in terms of electricity consumed, carbon footprint and global power consumption:

The main reason for this high energy consumption is that in the process of evaluating the proof-of-work algorithm, all other computers in the network are trying to find the solution. So, to solve the cryptographic puzzle, all the computers are free to participate but only the one which finds the solution will be rewarded with some bitcoins. So, only one computer will come up with the actual solution, and it then shares the result with all other computers in the network. This means that energy is consumed not only by the winning computer but also by all other computers in

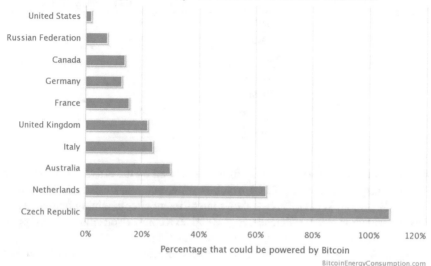

Fig. 5 Statistics of energy consumption of bitcoin [2]

Table 1 Energy consumption of a bitcoin

Description	Values
Current estimated annual electricity consumption (TWh) current minimum	73.12
electricity consumption (TWh)	57.76
Annual carbon footprint (kt of co_2)	35.830
Electricity consumed per transaction (kWh)	892
Carbon footprint per transaction (kg of co_2)	437.26
Minimum global power consumption of software (TWh)	22
Peak power usage of bitcoin network (TWh)	67

the network who are trying to find the solution. This process of finding the solution actually uses a lot of energy to repeatedly change and find a nonce value that matches the target. From the environmental point of view, a large source of electricity generation is the non-renewable sources of energy which are responsible for increasing the carbon footprint. It has also been found that blockchains use energy that is approximately equal to the energy consumption of a nation annually because of this process of mining. According to [10], it has been found that dishwasher energy consumption for a year is equivalent to the energy consumed by bitcoin network per transaction.

Bitcoins are mined using special mining hardware which were designed and improved over time to reduce their energy consumption. Earlier, CPUs were used to mine blocks which were slow and used more power. So, GPUs are used now, which calculates nearly 100 times faster than possible with a CPU and also uses

Table 2 Bitcoin mining hardware

Name	Power efficiency (W/Gh)
AntMinerS7	0.25
AntMiner S8	0.98
AntMiner S8	0.98
Avalon6	0.29
Antminer V2	1.0
BPMC red fury USB	0.96
Gekko science	0.33

less energy comparatively. This was further improved by the arrival of Application Specific Integrated Circuit (ASIC) which is faster and consumes less power than FPGA, CPU or GPU [11]. Table 2 shows the different bitcoin mining hardware and the corresponding power efficiency. Figure 6 gives a comparison among the different mining hardware on the basis of power efficiency (Fig. 7).

Figure 8 displays one of the world's largest bitcoin mines which is located in the industrial park of SanShangLiang, the outskirts of Ordos city, Mongolia. It is around 400 miles from Beijing, capital of China. As bitcoin mining consumes huge

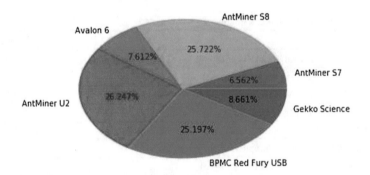

Fig. 6 Comparison of power efficiency of mining hardware

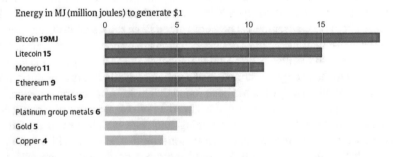

Fig. 7 Comparison chart of different energy consumptions [8]

Fig. 8 Bitcoin mining machine [12]

enormous amounts of electricity, miners found locations that offer cheap energy. Founded in 2014, Ordos mine is China's oldest large-scale bitcoin mining facility. It was acquired by Bitmain in 2015. It is powered by electricity mostly from coal-fired power plants. Its daily electricity bill amounts to 39,000 USD [12].

4 Methods to Reduce Energy Consumption in Blockchain

One way of reducing the energy consumption is by lightning network, where the transactions occur off-chain by positioning channels between the users and permitting the transactions to be recorded on closing of the channels [13]. It is used for fast transactions among the parties by directly setting up channels across them. In this network, channels will be setup between users, and thus the transaction will only be recorded on the blockchain when the channel is closed. But this would not lead to efficient solution for reducing energy consumption as very small amount of energy is used by the nodes to record transactions. The scheme was proposed by Thaddeus Dryja and Joseph Poon in 2015 [14]. Here, a user-initiated channel is designed which permits dealing with transactions without the engagement of any third parties. Information is stored only when the channels are closed. It is not stored on open state. Two parties are provided with a wallet differently (displaying the amount of bitcoin) on start of the transaction. First, the address of the wallet is saved, and then the virtual channel of the two parties is stored in blockchain. Every transaction is updated only when the whole transaction is done or completed [14]. This has minimized the energy consumption per transaction and hence consumes much less energy. Though a bit of energy has been reduced by faster transactions using lightning network method, the huge energy consumption for mining could not be minimized by this method. So, instead of using proof-of-work mechanism another algorithm called proof-of-stake can be used. PoS uses economic game theory in order to maintain network consensus. In this system, the network validators must deposit and lock-up or stake the coins to the network. In case of any fake transaction or any unfair means, the staked coins will be lost entirely. In PoS, the amount of coins staked, along with the amount of time, the coins that have been staked in the network will work as the parameters for choosing which validator will likely to be given chance to validate the next block of transactions. The validator will earn additional coins as a reward for its validation work. Thus, PoS becomes an energy-saving alternative of PoW, and this is the reason why bitcoin Oils PoS-based technology will be much more beneficial as compared with bitcoins current PoW model.

4.1 Proof-of-Stake

This protocol uses less computations and can be used for Ethereum and certain altcoins [3]. It was first implemented in a cryptocurrency named Peercoin [15]. According to this protocol, the creator of the next block is chosen randomly on the basis of wealth or age [4]. This means that miners with large amount of digital currencies can add a new block to the blockchain. In this case, not all computers are allowed to participate in the mining process instead only one computer is participating in the mining process. This saves the power that all the computers in a network waste searching for the solution in proof-of-work algorithm and cuts down the cost of mining by 99% [16]. So, proof-of-stake proves to be more energy efficient and cost effective than proof-of-work algorithm, and this reduces the need to release too many coins for keeping the miners motivated [15].

A modification of this protocol is the Delegated Proof-of-Stake protocol commonly known as DPos. Unlike the proof-of-stake protocol, where a user puts his coins on stake for acquiring the right to validate a transaction, DPos protocol allows users to vote a witness and the witness who gets maximum vote will get the right to validate a transaction. This protocol is also found to consume less energy as compared to proof-of-work and is also better than proof-of-stake [17] (Table 3).

4.2 Proof-of-Authority

In Proof-of-Authority protocol, there is a small group of people who validate the transactions and put their reputations at stake for validating a transaction. This protocol is found to be fast for doing computations with very less consumption of energy compared to proof-of-work [18].

Table 3 Comparison of mining protocols

Protocol	Characteristics
Proof-of-Work (PoW)	1. All users participate
	2. More energy consumption
	3. Slower than PoS
Proof-of-Stake (PoS)	1. User having more wealth participate
	2. Less Energy consumption
	3. Faster than PoW
Delegated Proof-of-Stake (DPoS)	1. Few users participate
	2. Less energy consumption than PoW
	3. Faster than PoS and PoW

Comparatively, it can be seen that proof-of-work uses the maximum energy. So, switching to other protocols like proof-of-stake or DPos or Proof-of-Authority can help in reducing the energy consumption for validating blocks or transactions.

4.3 Renewable Sources of Power Generation

In reality, the best solution for getting rid of such energy consumption problem is using renewable energy for mining of crypto currencies. Inner Mongolia, which is one of the main places associated with bitcoin mining, uses coal power for the mining procedure, and currently, more than 70% of bitcoin mining is done there only [19]. This inevitably leads to global warming along with pollution of nature. Countries like Iceland and Norway produce more than 90% of their energy from renewable resources. If mining is located in such countries, then making use of renewable resources will be much easier because they have plenty of these resources [19].

As blockchains use enormous amount of energy, instead of using the non-renewable sources of energy, the renewable sources of energy can be preferred for production of power. Renewable sources of energy are found to cause less pollution and will not increase the carbon footprint. Renewable sources of energy include solar, wind, water, etc. Companies like IBM and Intel are preferring green blockchains for the transaction process [20]. Also, blockchains can themselves provide the solution for power generation. Decentralisation of power is the first step towards it. Power will be generated by small communities using wind or solar energy [20]. That means the world of passive energy consumers is getting replaced by new consumers who are not only buying power but also selling it. Then comes, the use of blockchains. Blockchains will create peer-to-peer communications among the different consumers of power. This entire setup is known as microgrid [21]. Here, the consumers will pay using blockchains, and they have to pay each other for generating power. Using blockchains, power will flow from those producing in surplus quantities to those who are in need of power without any human intervention. This is how green blockchains can be used with decentralisation of power that is renewable [21]. One of such microgrids is set up in Switzerland that is named as MyBit. It is a decentralised energy grid. It uses Iot, artificial intelligence and solar energy and has combined the three to become a source of energy [22]. In UK, a startup named as Electron has been set up that uses blockchain technology for energy trading [22]. The main idea behind switching to green blockchains is promoting sustainable development which means that apart from meeting the needs of the present generation, we need to think about the future generation.

Reducing the energy consumption of the blockchains is the need of the hour. If these challenges can be overcome, then blockchains will become the best option for the future world. "If blockchain could provide an accounting system, it could turn the national grid from being the enemy of microgrids to being their friend"—Philip Sandwell, Imperial College London [23].

5 Conclusion

Blockchains are designed to provide security to all types of data so that no one can tamper with them. But also, at the same time, they are found to be consuming enormous amount of energy. So, various organisations and companies are trying to find the solution for solving the energy issues of blockchains, and at the same time, they are trying to switch to renewable sources of energy for sustainable development and for reducing any environmental hazard that is associated with the use of blockchains. Blockchains are very useful if implemented, but to use it in every field of the economy the workers and the employees need to be given proper knowledge of these blockchains. So, education is the main factor for the implementation of blockchains.

References

1. Bitcoin: A peer-to-peer electronic cash system by Satoshi Nakamoto. Available online: https://bitcoin.org/bitcoin.pdf
2. de Vries, A.: Bitcoin energy consumption index. Digiconomist (2014). Available Online: https://digiconomist.net/bitcoin-energy-consumption
3. Aras, S.T., Kulkarni, V.: Blockchain and its applications: a detailed survey. Int. J. Comput. Appl. 0975–8887 (2017)
4. Zheng, Z., Xie, S., Dai, H.-N., Chen, X., Wang, H.: Blockchain challenges and opportunities: a survey (2018)
5. Khan, I.: What is blockchain technology? A step-by-step guide for beginners. Blockgeeks (2016). Available Online: https://blockgeeks.com/guides/what-is-blockchain-technology
6. Huckle, S.: Bitcoins energy consumption is a concern but it may be a price worth paying; 7 Nov 2018. Available Online: http://theconversation.com/bitcoins-high-energy-consumption-is-a-concern-but-it-may-be-a-price-worth-paying-106282

7. Hern, A.: Bitcoin mining consumes more electricity a year than Ireland. 27 Nov 2017. Available online: https://www.theguardian.com/technology/2017/nov/27/bitcoin-mining-consumes-electricity-ireland

8. Hern, A.: Energy cost of mining more than twice that of copper or gold. The Guardian; 5 Nov 2018. Available online: https://www.theguardian.com/technology/2018/nov/05/energy-cost-of-mining-bitcoin-more-than-twice-that-of-copper-or-gold

9. Flipo, F.: The bitcoin and blockchain: energy hogs. 17 May 2017. Available online: https://theconversation.com/the-bitcoin-and-blockchain-energy-hogs-77761

10. Massessi, D.: Public versus private blockchain in a nutshell. Medium. 12 Dec 2018. Available online: https://medium.com/coinmonks/public-vs-private-blockchain-in-a-nutshell-c9fe284fa39f

11. Bitcoinmining (2010). Available online: https://www.bitcoinmining.com

12. Wong, J.I., Simon, J.: Photos: inside one of the world's largest bitcoin mines. 18 Aug 2017. Available online: https://qz.com/1055126/photos-china-has-one-of-worlds-largest-bitcoin-mines

13. Copeland, T.: How will Bitcoin solve its energy consumption problem? Newsbtc; 21 May 2018. Available online: https://www.newsbtc.com/2018/05/21/will-bitcoin-solve-energy-consumption-problem

14. Acheson, N.: What is Bitcoins lightning network; Coindes; 21 March 2018. Available Online: https://www.coindesk.com/information/what-is-the-lightning-network

15. Lisk Academy; Proof of Stake. Lisk (2018). Available online: https://lisk.io/academy/blockchain-basics/how-does-blockchain-work/proof-of-stake

16. Proof-of-Stake. Wikipedia; Available online: https://en.wikipedia.org/wiki/Proof- of-stake

17. Ray, S.: What is delegated proof-of-stake, 15 April 2018. Available Online: https://hackernoon.com/what-is-delegated-proof-of-stake-897a2f0558f9

18. Matthews, K.: 4 ways to counter blockchain's energy consumption pitfall; 18 April 2019. Available online: https://www.greenbiz.com/article/4-ways-counter-blockchains-energy-consumption-pitfall

19. Quora; Is Bitcoins energy consumption a problem for the world? 2018. Available Online: https://www.quora.com

20. Zhao, H.: Bitcoin and blockchain consume an exorbitant amount of energy, Feb 2018. Available online: https://www.cnbc.com/2018/02/23/bitcoin-blockchain-consumes-a-lot-of-energy-engineers-changing-that.html

21. Chebbo, M.: Powering a sustainable future: How blockchain can solve bitcoins energy consumption crisis. Itproportal; 01 June 2018. Available On-line: https://www.itproportal.com/features/powering-a-sustainable-future-how-blockchain-can-solve-bitcoins-energy-consumption-crisis

22. Voshmgir, S.: Blockchain and Sustainability. Medium; 11 Aug 2018

23. Baraniuk, C.: Microgrids and the Blockchain are powering our energy future. Wired; 12 Oct 2017. Available online: https://www.wired.co.uk/article/microgrids-wired-energy

Study on Network Scanning Using Machine Learning-Based Methods

Indranil Roy, Shekhar Sonthalia, Trideep Mandal, Animesh Kairi and Mohuya Chakraborty

Abstract Network scanning is among the first steps to determine security status of a computer network. Although there are many existing tools for scanning a network, they lack a key component—versatility. In the present day, there are multitudinous attacks that a network may be exposed to. Existing network scanning tools can scan for only those vulnerabilities that the scanner was designed to scan for. They lack the ability to efficiently adapt to newer threats. In this paper, we discuss the ways in which machine learning-based methods can improve accuracy and precision of network scanning. We also describe the approach we have adopted to implement this technique.

Keywords Machine learning · Network scanning · Intrusion detection · Vulnerability analysis

1 Introduction

Network security comprises of methods used for protecting a network against threats that may compromise the network's functionality, availability or allow unauthorized access or misuse of network-accessible resources. Network scanning is first line of

I. Roy · S. Sonthalia · T. Mandal (✉) · A. Kairi · M. Chakraborty
Department of Information Technology, Institute of Engineering and Management,
Kolkata 700091, India
e-mail: mandal.trideep1435@gmail.com

I. Roy
e-mail: indranil.r5198@gmail.com

S. Sonthalia
e-mail: sonthalia1996@outlook.com

A. Kairi
e-mail: animesh.kairi@iemcal.com

M. Chakraborty
e-mail: mohuyacb@iemcal.com

© Springer Nature Singapore Pte Ltd. 2020
M. Chakraborty et al. (eds.), *Proceedings of International Ethical Hacking Conference 2019*, Advances in Intelligent Systems and Computing 1065, https://doi.org/10.1007/978-981-15-0361-0_6

security measure that allows one to determine the hosts in the network and their related metadata like a domain name system (DNS), its IP address range, information about the subnet or private IP addresses that could be accessed remotely, their network structure, their operating systems and the services running on each host in the network, their network user and group names, routing tables, Simple Network Management Protocol (SNMP) data, etc. Network scanning includes both network port scanning and vulnerability scanning [1].

The structure of the paper is as follows. After the introduction in Sect. 1, brief overview of network scanning and machine learning has been described in Sect. 2. Section 3 contains description of a machine learning model for network scanning. Section 4 describes our application of the above-mentioned concepts. Section 5 consists of details of performance analysis that were performed. Section 6 concludes the paper with description of future work on this topic.

2 Network Scanning and Machine Learning

2.1 Network Scanning

Network scanning refers to mapping a network based on the hosts connected to the network, the services used by them, the type of communication and data that is sent among the hosts and to the Internet. Network can be scanned for a particular purpose like vulnerability scan or to maintain a record of the current state of the network. Port scanning is a part of network scanning where each host is probed for open ports. This is done to determine the services running on each port of each host [2]. Vulnerability scan helps identify weak points in the network which may be prone to internal or external exploitation.

2.2 Machine Learning

Machine learning can be defined as study of statistical models and algorithms that are used to make a computer system perform a specific task without being explicitly programmed to do so [3]. It does so by generating a mathematical model that is a representation of a set of data called 'training data'. This is a labeled data which is used to develop the mathematical model. Once the model is generated, new data known as 'test data' is given as input to the model. The outputs obtained are used to determine accuracy and precision of the model. A machine learning model has the advantage of processing large datasets with precision and accuracy.

3   A Machine Learning Model

The different stages and segments of a machine learning model are described below. Each individual stage is important for the overall functionality of the model itself [4].

3.1   Machine Learning Model Description

Problem Statement: Proper understanding of the problem statement is important for designing a network that can be efficiently used for learning and predicting data. Flaws in understanding of problem statement might result in development of an inefficient machine learning model.

Data Gathering: The aim of this stage is to gather large amounts of data with enough variation to eliminate bias. To obtain this data, the network is first mapped and data like packet headers, session logs, traffic details and performance metrics are recorded. Specific parts of the data are then extracted according to requirement and used to design the dataset. This dataset should contain both labeled historical data from logs as well as real-time network state and performance metrics. Data collected in real time also acts as feedback signals for the machine learning model.

Data Analysis: The dataset designed in the previous stage needs further processing before it is used by the machine learning model. To improve the training phase of our model, we need to clean the data and extract the features that have the most impact on the variations in the data (feature engineering). To clean the data, concepts such as discretization, missing-value completion and normalization are used. This is followed by feature engineering process. Since extracting features often need knowledge and insights into the network itself, the overall process is often difficult and time consuming. In such cases, this stage can be automated using a deep-learning model.

Model development: Model development consists of two phases—model construction and model validation. Construction of model includes selection of type of model, training the model and tuning the model. The learning algorithm is chosen based on its applications and other factors.

Model construction involves model selection, model training and model tuning. A mathematical model or algorithm needs to be chosen according to the size of the dataset, typical parameters of a network setup, the features available, etc.

Model validation is used to determine whether the learning algorithm can determine the parameters required for the mathematical model to be accurate enough to represent our data. Cross validation is generally used to determine the accuracy of the model and whether the model is overfitting or under-fitting. This provides insights on how to optimize the model further.

Deployment and Inference: When the machine learning model is finally deployed in a network, a few factors should be considered during the process. Factors like availability of computational resources, limits on response time of the model, definite preference among over-head and accuracy too are essential in determining real-world performance and stability of the model.

3.2 Areas of Application of Machine Learning

- Intrusion Detection—Intrusion detection aims to identify unauthorized use by users both outside and inside the system. A typical intrusion detection system uses statistical methods to detect an intruder, but machine learning can help us predict events of intrusion and even report intrusion attempts in real time.
- Network Cognition—Network cognition is inspired from cognitive actions of human brain. This aims to have an 'intelligent' network that can predict actions happening in the network before they can occur. This is where machine learning assists in analyzing the network and predicting the occurrence of events with accuracy.
- Traffic Prediction and Classification—A network may be designed to handle a certain type of traffic, but when deployed, there may be various types of traffic flowing through the network. Machine learning models can help classify the traffic for us to have an idea of what is flowing through the network. It can even be used to predict the amount of traffic when provided with a few parameters, allowing us make required modifications to handle the predicted traffic [5].
- Rogue User Detection—A machine learning model can individually analyze each user in a network to determine who among the users are exploiting their access to the network for their own benefit. This cannot be done manually as monitoring each user in real time is not humanly possible.
- Rouge Host Detection—Just like users who exploit their access to a network, there might be host systems that are vulnerable to exploit from outside. This could allow an outsider gain access to the network. Such host systems can be detected in real time with high accuracy using machine learning models.
- Persistence Programs Detection—A machine learning model that is scanning the entire network can detect programs that are maintaining constant connection with a host outside the network. Such a program has a high possibility of being a part of a persistence attack. Detection of such programs reduces the probability of the network being compromised by an attacker.
- Automated Resource Management—A machine learning model that is mapping the entire network can assist in resource allocation and sharing in real time in accordance with the traffic and number of clients accessing the network. Resource management can be done manually, but it would result in improper resource allocation as a human cannot allocate or claim back resources dynamically as the load on the network changes. A machine learning model provides this quick response and dynamic behavior, resulting in efficient resource utilization.

4 Our Approach—Machine Learning and Network Scanning

We have designed a behavior-based intrusion detection system that combines both machine learning and network scanning to obtain accurate results. The intrusion detection system is designed to run on a remote portable processor or Raspberry Pi model 3B. We have chosen to use a Raspberry Pi as it is a capable microprocessor platform that can handle network scanning tasks.

We have configured the single-board computer such that it runs Kali Linux out of the box and boots directly onto the home screen removing the issue of logging in every time on startup. This removes the problem of getting our Raspberry Pi stuck on the login screen every time we boot it up because when deployed at a client site it is hard to have our RPi connected to the Wi-Fi in order to SSH in the connection. Having our RPi directly jump onto the main screen with the script running on startup is a much easier process and eases the workload of the penetration tester and network administrator.

We have designed the whole system keeping one thing in mind and that is portability, so that any penetration tester can carry our model out in the field, switch it on, connect it to the Wi-Fi and start the process of intrusion detection. We have used SSL protocol for communicating with the RPi to issue commands and retrieve results.

In order to initiate the process of penetration testing, we need to be on the network as the Raspberry Pi. Since this device is going to perform a full analysis of the network it is connected to, we assume that we have been given necessary authorization by the network administrator before carrying out the penetration testing. Additionally, preventing remote connections to this device eliminates chances of the network that is being analyzed to be targeted by an unauthorized external entity.

The Raspberry Pi is connected to the network with an Ethernet cable. Once the module has been successfully connected to the network, the custom-designed script then initiates SSH connection so that the module can be remotely and securely controlled.

Some of the most used tools for penetration testing are listed below. We went through each of the tools to determine which of the tools can provide maximum functionality and operational efficiency along with cross-platform usability (Table 1).

The custom script is written in Python using different modules and libraries like autosploit, metasploit, nmap and burpsuite. Shell scripting is used to issue commands to the Linux system and execute the required Python scripts. This also allows us to format the output of various commands and use them as inputs to our Python scripts. The code designed is implemented using menu-driven approach which provides faster and intuitive access to all the different modules.

We have added port and network scanning including subnet network scanning, remote packet injection. Vulnerability analysis can also be performed but before that we have to banner grab all the ports in order to know about the protocols running on those open and partially open ports. Features like details of the port that are open,

Table 1 Few popular penetration testing tools [6]

Name of tool	Purpose	Platform
Nmap [7]	• Network scanning	Linux, Windows, FreeBSD, Solaris, macOS, SunOS
	• Host discovery	
	• OS detection	
Metasploit Framework [8, 9]	• Vulnerability testing	All versions of Unix and Windows
	• Exploit development and execution	
Hping [10]	• Port scanning	Linux, Windows, FreeBSD, Solaris, MacOS, Sun OS
	• OS detection	
Xprobe [11]	• Active OS detection	Linux
	• TCP fingerprinting	
	• Port scanning	
p0f [12]	• OS detection	Linux, Windows, FreeBSD, Solaris, macOS, SunOS
	• Firewall detection	
Httprint [13]	• Web server fingerprinting	Linux, macOS, FreeBSD, Win32 (both command line and GUI)
	• Web-enabled devices detection	
	• SSL detection	
Nessus [14]	• Vulnerability detection	Linux, Windows, FreeBSD, Solaris, macOS, SunOS
	• Denial of service and misconfiguration analysis	
Shadow Security Scanner [15]	• Network vulnerability detection, audit proxy and LDAP servers detection	All Windows Server, Business and Professional Versions
İss Scanner [16]	• Network vulnerability detection	Windows Professional and Server Versions
GFI LAN Guard [17]	• Network vulnerability detection	All Windows Server, Business and Professional Versions
Brutus [18]	• FTP, Telnet and HTTP password brute-force attack	Windows 9x/NT/2000

the protocols running on those ports, the encryption used by those protocols and the versions of the services running are used as test data for our machine learning model.

We have used a logistic regression model as it would help us to predict the probability of a host being used for intrusion into a network. We have used the KDD-cup dataset to train our model using supervised learning. KDD-cup dataset is a popular dataset used for anomaly detection. This dataset was also popularly used for detecting novel attacks on networks [19]. We have used 17 features out of 41 total features in this dataset for our logistic regressor. We then trained our model on the most ideal and secure parameters that should actually be used in a network. This allows us to detect vulnerabilities in the network efficiently. Additionally, our system regularly scans the network for occurrence of any changes in its configuration. If a change is detected, it again obtains the new parameters and uses the logistic regressor to determine whether the system would become vulnerable to threats or not.

To improve the performance of the machine learning model, the parameters of the module can also be modified to better suit the requirements of a specific network, which would be mentioned by the network administrator. This would improve detection rates and accuracy of the model while reducing time required for the model to adapt to the particular network.

5 Performance Analysis

We have tested our system in comparison to a standard desktop running Kali Linux. We had created model networks with various network devices and different hosts. We then set up both systems to start mapping the network and fingerprint the hosts. Both the systems, our module running on Raspberry Pi and our script running on the desktop, returned the same results in almost similar time duration. We had repeated this for multiple iterations, and there was no significant deviation from this trend. We also simulated intrusion using one of the connected hosts to deploy payloads to another host using Metasploit. Both our module and the desktop running our script were able to report intrusion into the network within similar time duration.

The results we had obtained were that the performance of our system is similar to that of a standard desktop. We had finally generated a report of the network that contained details of the network like the connected hosts, the services running on them and their versions. The report also contained details of the hosts that are vulnerable to intrusion. During our simulated intrusion attack, the report also contained details of the host that was used for gaining access into the network and the common fixes like version upgrades that can be done to improve the security of the network.

6 Conclusion

Our Raspberry Pi-based module is an example of how concepts like machine learning and network scanning can be combined together and placed on a mobile platform. This ensures balance between performance and flexible deployment. We have developed this module with purely academic interest in our mind. Currently, the module can perform vulnerability analysis and intrusion detection. We plan to expand this unique menu-driven approach to include a suite of functionalities for complete penetration testing and report generation.

References

1. Gupta, A., Klavinsky, T., Laliberte, S.: Security through penetration testing: internet penetration. InformIT. Pearson PLC. Retrieved 2013-03-31 (2002)
2. RFC 2828 Internet Security Glossary
3. Mitchell, T., Buchanan, B., DeJong, G., Dietterich, T., Rosenbloom, P., Waibel, A: Machine Learning. Annu. Rev. Comput. Sci. **4**:417–433 (Volume publication date June 1990)
4. Wang, M.., Cui, Y.., Wang, X.., Xiao, S., Jiang, J.: Machine learning for Networking: Workflow, Advances and Opportunities. IEEE Network https://doi.org/10.1109/mnet20121700200
5. Boutaba, R., Salahuddin, M.A., Limam, N., Ayoubi, S., Shahriar, N., Estrada-Solano, Felipe, Caicedo, O.M.: A comprehensive survey on machine learning for networking: evolution, applications and research opportunities. J. Internet Serv. Appl. **9**, 16 (2018)
6. Bacudio, A.G., Yuan, X., Chu, B.-T.B., Jones, M.: An overview of penetration testing. Int. J. Netw. Secur. Its Appl. (IJNSA) **3**(6) (2011)
7. Nmap—free security scanner for network explorer. http://nmap.org/. Accessed 23 Nov 2011
8. MetaSploit.: http://www.metasploit.com/. Accessed 23 Nov 2011
9. Skoudis, E.: Powerful payloads: the evolution of exploit frameworks (2005) http://searchsecurity.techtarget.com/news/1135581/Powerful-payloads-The-evolution-of-exploit-frameworks. Accessed 23 Nov 2011
10. Sanfilippo, S.: Hping—active network security tool. http://www.hping.org/, Accessed 23 Nov 2011
11. Xprobe2.: http://www.net-security.org/software.php?id=231. Accessed 23 Nov 2011
12. P0f.: http://www.net-security.org/software.php?id=164. Accessed 23 Nov 2011
13. Httprint.: http://net-square.com/httprint/. Accessed 23 Nov 2011
14. Nessus.: http://www.tenable.com/products/nessus. Accessed 23 Nov 2011
15. Shadow Security Scanner.: http://www.safety-lab.com/en/download.htm. Accessed 23 Nov 2011
16. Iss Scanner.: http://shareme.com/showtop/freeware/iss-scanner.html. Accessed 23 Nov 2011
17. GFI LAN guard.: http://www.gfi.com/network-security-vulnerability-scanner. Accessed 23 Nov 2011
18. Brutus.: http://download.cnet.com/Brutus/3000-2344_4-10455770.html. Accessed 23 Nov 2011
19. Tavallaee, M., Bagheri, E., Lu, W., Ghorban, A.A.: A detailed analysis of the KDD CUP 99 Data Set. In: Proceedings of the 2009 IEEE Symposium on Computational Intelligence in Security and Defence Applications (CISDA 2009)
20. Bishop, C.M.: Pattern Recognition and Machine Learning. Springer. (2006). ISBN 978-0-387-31073-2
21. Henry, K.M.: Penetration testing is the simulation of an attack on a system, network, piece of equipment or other facility, with the objective of proving how vulnerable that system or "target" would be to a real attack. Penetration testing: protecting networks and systems. IT Governance Ltd. (2012). ISBN 978-1-849-28371-7
22. Faircloth, J.: Chapter 1: Tools of the Trade. Penetration Tester's Open Source Toolkit, 3rd ed. Elsevier. (2011). ISBN 978-1597496278
23. Nmap license.: Retrieved 2019-01-21
24. Nmap.org. Nmap Scripting Engine: Introduction. Retrieved 2018-10-28
25. Lyon, G.F.: Nmap Network Scanning: The Official Nmap Project Guide to Network Discovery and Security Scanning. Insecure.com LLC. p. 468. (2009). ISBN 978-0-9799587-1-7
26. Haines, J., Ryder, D.K., Tinnel, L., Taylor, S.: Validation of sensor alert correlators. IEEE Secur. Priv. **99**(1):46–56 (2003). https://doi.org/10.1109/msecp.2003.1176995
27. Medeiros, J.P.S., Brito Jr., A.M., Pires, P.S.M.: Computational Intelligence in Security for Information Systems. Adv. Intell. Soft Comput. **63**, 1–8 (2009). https://doi.org/10.1007/978-3-642-04091-7_1. ISBN 978-3-642-04090-0
28. Metasploit.: Metasploit. www.exploit-db.com. Retrieved 2017-01-14

29. Foster, J.C., Liu, V.: Sockets, shellcode, porting and coding: reverse engineering exploits and tool coding for security professionals. Chapter 12: Writing Exploits III. ISBN 1-59749-005-9
30. Foreman, P.: Vulnerability Management. page 1. Taylor & Francis Group (2010). ISBN 978-1-4398-0150-5
31. Bishop, M., Bailey, D.: A critical analysis of vulnerability taxonomies. Technical Report CSE-96-11, Department of Computer Science at the University of California at Davis, September 1996
32. Kakareka, A.: 23. In: Vacca, J. (ed.) Computer and Information Security Handbook, p. 393. Morgan Kaufmann Publications. Elsevier Inc. (2009). ISBN 978-0-12-374354-1

Is Blockchain the Future of Supply Chain Management—A Review Paper

Debopam Roy, Debparno Roy, Dwaipayan Bhadra and Baisakhi Das

Abstract The basic concept of supply chain management (SCM) is to handle the flow of information, goods, and services in an efficient way to achieve better performance and minimize risk. In SCM, information or data is the key resource for collaboratory. However, it has been found that in many cases, there is lack of trust among collaborators. Blockchain technology is considered to facilitate security and privacy for several applications. This paper focuses on some case studies of SCM integration with blockchain and that how the technology allows transfer of transactions securely in a digital decentralized ledger in absence of any mediator.

Keywords Blockchain · Supply chain management · Security

1 Introduction

Blockchain is currently the blooming topic of interest in the field of technology [1, 2]. The main reason of its growing interest is that the applications work through trusted system and can be operated in a decentralized manner where there is no need for third-party verification [3]. This gave birth to trustless network because in blockchain, transfers can be made even without trusting each other. It first emerged in the context of bitcoin, where it serves as a decentralized, distributed digital ledger recording all bitcoin transactions [4]. Bitcoin is a currency that is controlled by the network of users instead of by centralized banks. Through the use of bitcoin, money can

D. Roy (✉) · D. Roy · D. Bhadra · B. Das
Department of Information Technology, Institute of Engineering and Management,
Salt Lake Electronics Complex, Sector V, Kolkata, India
e-mail: roydebopam16@gmail.com

D. Roy
e-mail: roydebparno@gmail.com

D. Bhadra
e-mail: dwaipayanbhadra1998@gmail.com

B. Das
e-mail: baisakhi.das@iemcal.com

be transferred directly [5]. Monetary transferring is done directly through bitcoins. Other applications regarding blockchain will be discussed in Sect. 3.

A supply chain management (SCM) is a system of multiple processes and subprocesses. The main concept is to handle the flow of information, goods, and services in an efficient way to achieve better performance and minimize risk. As we know that many collaborators are involved in the SCM processes, often there is lack of trust between them. Therefore, it is found that the ultimate goal of SCM is to work as a single unit, to produce, to design, and to develop processes in production of goods often faiths.

The process of managing the entire network becomes tedious [5]. As blockchain technology is cryptographically secure, this paper focuses on some case studies of SCM integration with blockchain technology and how the integration helps to transfer transaction securely.

This paper is organized as follows. Section 2 explains the structure of blockchain and its preliminaries. The various application areas of blockchain are discussed in Sect. 3. Section 4 explains the concept of supply chain management, how blockchain can be implemented in the supply chain management system, and what are the advantages we are getting by applying blockchain in supply chain management. The main focus of this paper is in the case studies part. Then, three applications of supply chain management are discussed, and comparison of the today's scenario of those applications with the scenario which we will be having after applying blockchain technology in it and how blockchain can solve the problem which are faced in today's scenario is also discussed.

2 Preliminaries of Blockchain

As stated in [6], a blockchain is a growing list of records, called blocks, which are linked using cryptography. Every block in the blockchain consists of the hash cryptography of the previous block as shown in Fig. 1 which shows that the hash of block 2, i.e., 8L7H, is linked with the hash of block 1, i.e., KBNN, so a chain is formed. So it is basically a decentralized database which helps in maintaining the growing list of records between multiple authoritative domains [7].

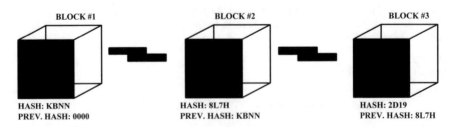

Fig. 1 Structure of blockchain

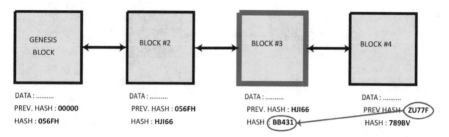

Fig. 2 Blockchain as an immutable ledger

2.1 Immutable Ledger

As stated earlier, all the blocks in the blockchain have cryptographical linkage between them. But what that actually means? So here comes the concept of immutable ledger. Suppose A wants to buy a house, he pays the money for the house, and in return he gets a title deed to the house which he takes to the Council in order to get register our ownership to the house. But how do they register? They write it down in a ledger. Some modern governments use digital ledger in this case. In some Third World countries, they even note down the ownership in a book written in hand. So if any attacker comes and messes up with that book or the house burn down along with that book, then technically his property does not belong to him anymore. Even with Excel spreadsheet, it will not be difficult for anyone to mess up with it. So what can he do? What if he introduces the concept of blockchain here? So anytime when someone buys a new house, a new block gets added to the blockchain. Suppose after a couple of months an attacker comes and decides to take his house away from him by tampering with the data in the blockchain. In those couple of months, many other property transactions took place and many new blocks were added to the blockchain. Now if the attacker tampers with the data in the block, then that will change the hash of the block and the cryptographic link will not work anymore because the hash of that block will not match with the previous hash of the next block [8].

So the attacker has to change every block which comes after our block which is technically impossible, and that is what we meant by saying immutable ledger which means once the data is entered in the blockchain it cannot be changed (Fig. 2) [8].

2.2 Distributed P2P Network

Distributed peer to peer (P2P) is the peer decentralized network of computers. Suppose the attacker was able to change the hash of the blocks. So how are they going to recover the lost data? Blockchain is a decentralized P2P network. It is a system where all the computers are interconnected with each other as shown in Fig. 3.

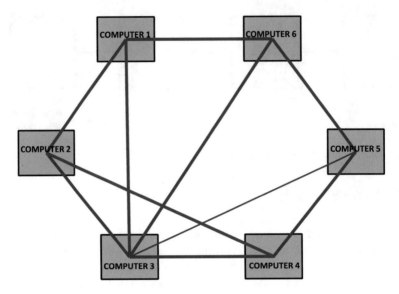

Fig. 3 Structure of distributed peer-to-peer network

When a new block is added, that information is communicated throughout the network and that block is added further throughout the network until all the computers have this block.

So the block that was created when A purchased the house was send to all the peers of the network. If an attacker comes to take away A's house, the peers are going to create a problem for them because when the hash of A's block is changed all the blocks after that are going to be invalidated. Even if they manage to change the blocks after that, the network is going to create problem. The system is designed in such a way that the network is always synchronized. The blockchain system is constantly validating the peer networks to check the malfunctioning. In case of any fault, the values from other computers will be copied over automatically and the original values will be restored [9]. So, it means the attacker cannot attack a particular computer in the blockchain. The attacker has to attack all the computers in the networks at the same time which is practically impossible.

2.3 Mining

Mining is the process of adding a new block to the blockchain, that is, validating a new transaction and registering them in global ledger, while mining the block number is assigned automatically. Once the data are mined in the blockchain, it cannot be changed. Miners have access to nonce part of the blockchain. Nonce is a random number which can be used only once in the blockchain. The nonce generates a hash which must meet the requirement in the blockchain. Suppose the hash requirement

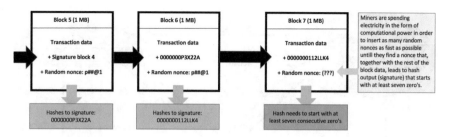

Fig. 4 Mining of block

of the blockchain is minimum seven consecutive 0's in beginning as shown in Fig. 4. Therefore, the job of miners is to enter the nonce (say p##@1) that generates the hash which matches the requirement of the blockchain [10].

So now if an attacker comes, all he can do is change the nonce of the block and replace it with another nonce which matches the requirement of the blockchain after lots of calculations and repeat the same process with all the blocks which comes after that [11].

2.4 Byzantine Fault Tolerance

What if more than one computer in the blockchain is attacked? What is the way to find out which computers are having the original copy of the blockchains and which computers are attacked? Here comes the concept of Byzantine Fault Tolerance. So what is it and how is it related to blockchain?

Figure 5 is the demonstration of the Byzantine Fault Tolerance. Byzantine Fault Tolerance is an algorithm to find out the traitor and take the actual action to attack or retreat. If all of them attack the castle together, they will win or if they all retreat together they will be safe. But one of them is a traitor. There must be one commander among them, and others are lieutenants. Suppose the commander told all the lieutenants to attack as shown in Fig. 5, but the traitor, i.e., lieutenant 3, told the other lieutenants that the commander told him to retreat. So in order to prevent chaos and to find out who is actually lying, they decided to interact among each other and find out that the commander actually told them to attack. So they all attack together and won the castle and also find out that lieutenant 3 is traitor [11].

2.5 Consensus Protocol

As many computers interconnected in the blockchain network, if the attackers come and change the hash of a block in the blockchain, the entire system will stop working because the blocks in that system will not be matching with the copy of the blocks

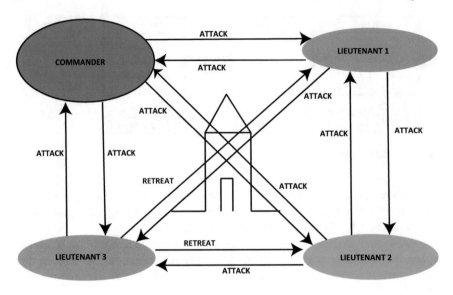

Fig. 5 Byzantine Fault Tolerance

in other computers in the system. But in order to find out in which computer the data are tampered, they verify with all the copies in the blockchain and take the majority into consideration. So in this way, the system recognizes the error in the blockchain and automatically replaces the data which is tampered with the data which is taken into consideration. This is the role of consensus mechanism in blockchain [12].

In Fig. 6, it is shown that in computer 2, an attack has added a new block copy of which is not shared by all other computers in the platform. Hence, the block

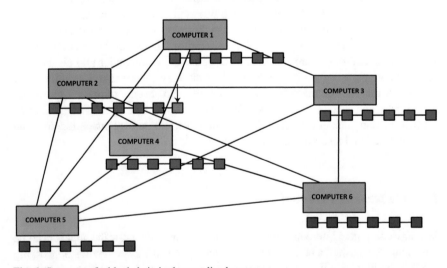

Fig. 6 Structure of a blockchain in decentralized system

is showing error (red colored). So ultimately that block will be removed from the system since it is verified that the red-colored block from computer 2 is not matching with other copies of the block in the blockchain system.

So from all the above discussions, we can conclude that the blockchain is a modern approach of organizing database; it keeps the record of every single change ever made in the system, and it arranges the data in a complete interconnected chain system.

3 Applications of Blockchain

- **Land registration**—For keeping details of the various descendants of land and also time stamping the details. Adding of any new descendant or altering land size adds a new block to the blockchain. These records are secure than the ledgers maintained by courts as that can be stolen, altered, or get misplaced or can even get burnt, and once this occurs, there remains no proof that the land belongs to whom. Blockchains if used in here it ensures that the information is safe and cannot be altered, also if cannot be obtained from one computer can be obtained from other computers. It is also not possible to tamper with the data [13].
- **Healthcare**—For keeping the records of the patients the kind of treatments they receive and the duration of their treatments. These records consist of data about medicines, report on affected areas of body, also images like X-ray, etc. The records are kept in the blockchain, and that cannot be tampered by the patients or the physicians. These records can be referred later by the authorities [13].
- **Voting system**—It can be used in voting system to prevent any tampering with the vote of the individuals. The vote of the people can be taken online, and their records will be kept in the blockchains; this prevents any tampering with the votes like once a person has voted that person cannot vote again also, he/she cannot change the vote again. So, this is more secure compared to ballot boxes [13].
- **Industries**—Used in various industries to store details of employees, various projects they worked on and other important information. Details of employees can include their salary information, when that was paid, when it will paid next, present working status of the employees, etc. It also contains details of projects like the name of the investor, the amount funded for that, the progress of the work, etc., and most importantly, these information cannot be changed once they are kept inside the blockchains. So, blockchains are preferred by several industries [13].
- **Cryptocurrencies**—There are various cryptocurrencies developed using blockchains, for example, Litecoin, Bitcoins, and Ethereum. The main idea behind using cryptocurrencies is that the physical currencies like the coins or the notes can be produced as duplicates. Currencies can even get stolen, and once lost or stolen, it is very difficult to retrieve them. So, switching to cryptocurrencies is thought as a better option. Cryptocurrencies are implemented using blockchains so it is not possible to produce fake cryptocurrencies and cannot be stolen in real sense and thus is more secured [13].

- **Supply chain management**—Supply Chain Management is called the process of management of the production of various goods and commodities which include all the processes of production starting from the initial processing of raw materials to the final processing of finished products. It is characterized of numerous business activities that occur at various levels of production of these goods, and each and every level is monitored by managers who make sure that the production unit maximizes its production capacity and yields the maximum output along with maximum profit and customer service. Details regarding this topic will be discussed in the following topics in details along with the case studies [14].

4 Supply Chain Management (SCM)

A supply chain is referred to as a network of various companies that work as a single unit to produce, develop, design, and service products involving various processes that finally result in production of goods. A network formed from the various facilities includes material flow from supplier and higher-level suppliers, transformation of raw materials into finished and semi-finished goods, distribution of final products among various customers, etc. The process of managing such a network of processes is termed as supply chain management.

A supply chain management (SCM) is a system of multiple processes and subprocesses that implement and try to make use of various function and other elements that are connected to a supply chain. The organization of a supply chain management system is based on numerous aspects that are correlated to one another. It involves various procedures that are necessary in transformation of goods into final products, and the people responsible for this job are very well connected to one another through activities like purchasing, delivery, packaging, forming numerous supply alliances that play a key role in the final distribution of the goods and products that are to be shipped by the distributors. Effective flow of information through supply chain management is very important for the continuation of all these processes in a systematic manner. Key business areas include customer service, enterprise performance, order management, and demand planning (Fig. 7) [5].

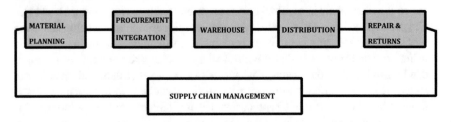

Fig. 7 Various stages in the SCM cycle

Pros/Benefits of supply chain management include reduction of uncertainty, proper inventory levels in the chain, elimination of unplanned activities, good customer services, minimization of delays, etc., whereas Cons/Disadvantages of supply chain management include delayed production and distribution, lack of coordination among business partners, poor quality of products, poor demand for products, various interferences with production, increased pressure on customer service unit, and making the entire system complex.

5   Use Cases

Although a variety of blockchain use cases already exist, not all of them seem to rely on blockchain specific features, but could rather be solved with traditional technologies. The purpose is, therefore, to identify characteristic use cases described for blockchain in the field of SCM and to analyze them regarding their mindful technology use based on five mindful technology adoption principles: engagement with the technology; technological novelty seeking; awareness of local context; cognizance of alternative technologies; and anticipation of technology alteration [15]. Implementation of the blockchain in the SCM would result in the following:

- Faster transaction between peer to peer with very few middlemen or intermediaries
- Transactions will be conducted simultaneously from both ends at the same time
- Automatic updating of ledgers
- Low cost
- Computing power will allow less use of manpower
- Transactions are authorized, encrypted, and accessible to participants with authorization and are also traceable within the ledger
- Transactions are immutable [16].

In this content, we have studied the case of three major areas where the integration of SCM with blockchain is advantageous.

5.1   Waste Disposal Management System

As stated in [17], waste management or waste disposal is all the activities and actions required to manage waste from its inception to its final disposal. This includes among other things, collection, transport, treatment, and disposal of waste together with monitoring and regulation. It also encompasses the legal and regulatory framework that relates to waste management encompassing guidance on recycling, etc. From Fig. 8, we can see that in today's world, waste disposal management is done mainly in six major steps:

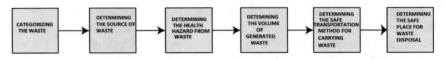

Fig. 8 Today's scenario of waste disposal management system

i.   Initially, the categorization of various solid wastes are done, i.e., whether the generated wastes are organic (food) or combustibles (paper, wood) or non-combustibles (metal, tin) or ashes/dust or bulky (tree brunches) or dead animals or hazardous (oil, battery, medicine) or constructive waste (broken concrete).

ii. Then determination of the source of waste is done, i.e., medical centers or food stores or feeding centers or food distribution points or slaughter areas or warehouses or agency premises or markets or domestic areas.

iii. After that the associative risks are also needed to be find out. Many transmitted diseases occur from organic wastes. Various infections and skin diseases occur through hazardous wastes.

iv. It is necessary to know the volume of waste generated in order to collect them in various size containers.

v.   For carrying wastes to the final disposal point, the transportation method is required. Transportation methods are mainly of human-powered or animal-powered or motor-powered type.

vi. Determining the safe place for waste disposal is the most important factor of waste management system. Wastes are generally carried to a large wasteland and disposed by land-filling, incinerating, composting, and recycling [18].

SCM integration with blockchain technology—Fig. 9 shows the blockchain integration with waste management system. Firstly, there will be an administrator who manages the entire waste management system in the blockchain. In this platform, the customers will be able to create their account in the blockchain and place their

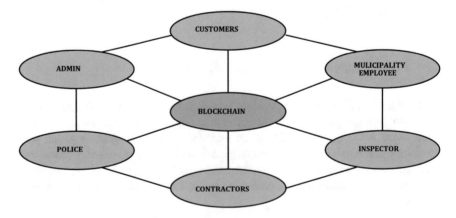

Fig. 9 Blockchain-enabled waste management system

request for collecting the wastes from their home. Every customers will be able to register to the blockchain with the help of necessary information like name, address, ID proof, etc. The customers will have access to the blockchain once the registration is approved by the administrator. After the registration is completed, the customer can perform the request for waste collection in the platform and attach a document regarding the waste type (industrial, organic, combustible, non-combustible, hazardous, etc.) and the location from where the waste will be collected. Daily basis customers can also sign contracts for the waste collection. Various online payment methods must be available along with cash-on-delivery option (Fig. 9) [19].

The inspector will be the one who will create the profile on the distributed peer-to-peer immutable ledger platform and will visit the waste disposal site for necessary inspections. He will decide which site will be used for waste management and upload the inspection details on the blockchain. He will also receive the feedback form from the customers and take necessary steps accordingly. The contractors will be able to create their account in the blockchain and perform the task assigned to them by the municipality. The contractor's job is to find out the quantity of waste collected, can the waste be recycled, whether the waste is hazardous or not, etc., and inform back the details to the municipality. The external entities like police will also be able to create their account on the platform. They are mainly hired by the municipality to investigate in case any accident has occurred and find the cause of the accident and report back to the municipality. There will be an admin who will be responsible for controlling everything in the blockchain platform. Their main roles will be managing all the registrations and authorization, managing all types of waste disposal services, and managing the payment and fines of everyone linked to the platform [19].

Remedies with Blockchain Technology

- **Fraudster and Report Manipulation**—While disposing the waste, the contractor submits their waste plan report to the municipal employee for receiving their wages. The wages are mainly based on the quantity of waste disposed, but in the current state there is no way of determining the quantity of waste disposed or verifying the report submitted by the contractor. So the report can be manipulated easily. But in blockchain, the data cannot be manipulated. So when it comes to paying the contractor, the waste disposal management activity history can be rechecked and the contractors will be paid accordingly.
- **Information Loss**—In today's scenario, all the documents regarding the waste management system is stored in a file. So there is a high probability of data loss when it comes to any sort of accidents like fire or of other sort or during file manipulation by fraudsters. In blockchain, the data are stored in the form of a distributed ledger. It means many computers in the platform will have the same copy of the document. So in case any change is noticed in a particular copy of the record in the platform, it will be immediately informed to the computer where the change is been noticed and will be replaced with the data of the other computers in the system. So information loss will become impossible via blockchain [19].

5.2 Food Supply Chain Management System

The food supply chain management is a major supply chain management system that has faced difficulties in modern times with its overall system. Various cases of food contamination and illness due to poor food quality are reported every year around the world. Recent studies have shown that one out of ten people suffer from various food contamination related illness [20].

There is a general concern among people whether to buy food supplies without knowing every little detail from the moment it is set to be shipped to the customer till the customer finally has the product safely delivered to them thus a need for transparency is created in this field [20]. Figure 10 shows the current scenario of food supply chain management system. The food production begins with the processing of raw materials and their conversion into finished products with intermediary processes like quality checking and approval. The finished products are then distributed which is followed by various processes like marketing and advertisement of products or in other cases directly sold at the market for consumption.

Sometimes these products are put up for online marketing which additionally involves a team that is devoted to receiving and sending purchase orders. The problem faced by these different units involved in this supply chain management is that

- These units perform their respective works individually, and none of these units stay connected to one another.
- Each one of these units act as a separate management system regardless of the fact that they are all part of a bigger supply chain management system.

In order to overcome this situation, the modern approach of blockchain implementation has been adopted.

Remedy using blockchain With the implementation of blockchain in the food supply chain management system shown in Fig. 11, this difficulty can be resolved.

- By the help of blockchain, we can see that the producers, suppliers, retailers, and finally the consumers are all connected digitally with one another, and all intermediate processes involved in the production till delivery of the food supplies are monitored by each and every person involved in the management process [21].
- The food production department can easily monitor the raw materials and supplies that are need for the production process, and with the help of blockchain, they can

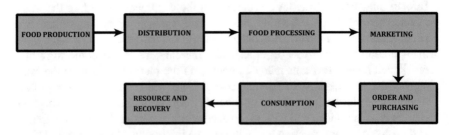

Fig. 10 Today's scenario of food supply chain management system

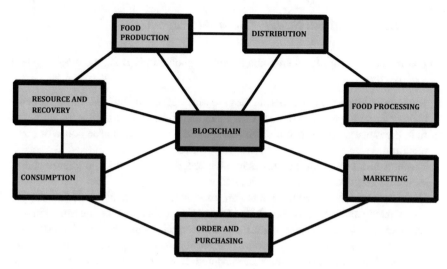

Fig. 11 Blockchain-enabled food supply management system

upload all required information in the blockchain database system from which all necessary data can be retrieved by any department involved in the production unit.

- The suppliers can keep track of all information about their customer's orders and cancellations by their respective IDs that they provide in the platform, and any enquiry or complaints or any kind of feedback that is received are stored for future purposes.

- The retailers provide all information about the current market status and price of each food product that is being sold and that helps both the customer and the management system to stay updated and thus helps the overall production unit.

- The platform that is created by the help of blockchain benefits the consumer as all kinds of customer-friendly services are made available like online ordering and shopping, shipment tracking, emergency cancellation, and special customer services like food quality complaint department, and food tampering and damage control department which helps the customer be more secured about the food products they are consuming, and it also helps to develop a better understanding between the customers and the producers.

The production of food products and also their transportation are made traceable for the consumers so that they know exactly what the problem that their product is facing, and if such a case arises, then they can know about the mishandling, contamination, or expire of food products before buying it. This blockchain implementation in food supply chain provides benefits not only to the consumers, but also to the producers, suppliers, and retailers [21].

5.3 Pharmaceutical Supply Chain Management

In pharmaceutical supply chain management as shown in Fig. 12, the whole system works like this.

- **Raw materials**: Raw materials are the essential ingredients such as different chemicals or minerals that are required for manufacturing of a particular medicine; at first, the raw materials are collected and supplied to the production house or manufacturer.
- **Production**: The medicine gets manufactured, and afterward, the manufactured medicine gets transferred to the distributors.
- **Distributors**: The distributors have an important role in pharmaceutical supply chain management because they help in managing a big amount of chemical inventory that can exceed the drug manufacturers space allowance for the chemical hazard classes. Distributors can outsource different parts of the pharmaceutical supply chain such as sampling of raw materials and cGMP released goods storage, and from distributors it reaches different pharmaceutical companies or hospitals.
- **Pharmacy**: And finally from pharmacy or hospitals, those medicines get consumed by the end patients; till now the whole system works like that [22].

In this kind of present supply chain scenario, if anywhere in the supply chain this different types of fraud takes place at different places in the supply chain, that may be in the producer section or in the distributor section, it will be very much difficult for the customers or the pharmaceutical companies to detect that, actually where the fraud took place. So from the above shown scenario, it can be well understood that there is a big transparency gap in the whole system. Because of this present supply chain system, the black marketers are getting a huge opportunity to counterfeit the drugs, which is a great hazard for the human health. These counterfeited drugs are being distributed to the customers without getting traced or detected. By the distribution of this kind of drugs, the developing countries are getting badly affected where counterfeit drugs are about 10–30% of the total.

Therefore, it is becoming quite important for the suppliers and the pharmaceutical companies to undertake such a digital technology with the help of which it will be quite easier to detect if any fraud is taking place or not [23]. Basically, the counterfeiting occurs at the manufacturing department or it is done by the distributors. This counterfeiting of drug usually takes place by entering counterfeit drug as authentic in the supply chain [24]. Because of all this things, the customers are losing trust in the present supply chain system. Because the blockchain technology has got an immutable nature, it is very helpful to trace drugs as it goes from manufacturer to different customers. By using the blockchain technology, anyone who is involved in

Fig. 12 Today's scenario of pharmaceutical supply chain management

the supply chain will be able to check whether the system is compromised at any place [7, 24, 25].

Remedy using blockchain Now due to the increase of number of customers and increasing number of scams, there needs to be a reliable system upon which the stockholders can trust; in this situation, blockchain comes in the scenario. Blockchain will be very helpful to solve this problem because it is becoming very popular in terms of application in supply chain management. System vulnerability in the drug supply chain leads to many pain points such as very little visibility for tracking and authenticating the product. By introducing blockchain, many such problems can be solved very easily. Customers will be able to tag the drug with the help of barcode, and if any kind of scam takes place, the records can be kept on the blockchain in the form of secure digital block; these records will keep on updating time to time as the drugs will get carried or transferred from one particular place to another in the supply chain. The persons who have an authorized access including the patients or the end customers will be able to check or track the movement of drug any time [23].

Due to the immutable nature of the blockchain, it provides drug traceability from manufacturer to the customers, and by the usage of the blockchain the patients or any authorized user will be able to check the system if the system is compromised anywhere. Apart from ensuring product integrity and counterfeiting efforts, blockchain also helps financial development [23].

Blockchain along with the help of IoT device will be very helpful in tracking the condition and position of the drug for the stakeholders, where the full provenance of a unit, its condition, authority rights, and checkpoint approval could be accessed at any point of time. If there any deviation occurs in the condition of the drug such as the temperature condition of the drug, it will get instantly tracked by the IoT device whose data will be the input and can be detected by the help of smart contract on the blockchain, then notifications will be executed by the smart contract rules to take effective actions by the stakeholder in charge of that particular phase of the supply chain. With the help of blockchain, every location of the drug, every activity with that drug can be traced by the end user, and thus the transparency problem can be solved; each and every person in the chain will have his/her personal ledger, in which the data will be kept updating time to time; therefore, if any tampering takes place, it will be instantly captured and required steps will be taken. So with the help of blockchain, we can get a fully secure pharmaceutical supply chain management system [24].

From Fig. 13, it can be seen that all the parties those who are involved in the whole supply chain are connected with each other with the help of blockchain, transferring of products or raw material from one person to another person gets immediately updated in the blockchain. If any kind of problem arises anywhere in the whole system, it will get immediately detected in blockchain system. Therefore, by using blockchain technology, the end customers will get continuously updated about the condition and position of the product, further with the help of blockchain technology the customers will be able to trust the supply chain which will be of much help for the whole business as well as the health of the customers [25].

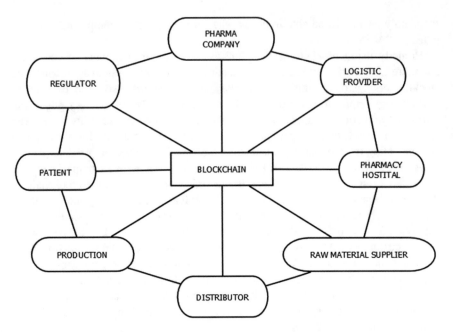

Fig. 13 Blockchain-enabled pharmaceutical supply chain management

6 Conclusion

SCM integration with blockchain is the main focus of this paper. In this paper, the problem that has been faced in today's world in supply chain management and how blockchain introduction puts an end to the need of trusted third parties for both data requirements and supply chain complexity is described in this paper [26].

Moreover, with the help of blockchain technology, making legal and regulatory decisions regarding collection, storage, and sharing of important data are made much simpler. Moreover, laws and regulations are programmed into the blockchain itself, so that they are enforced automatically [26].

References

1. Announcing the Secure Hash Standard: Available online: http://csrc.nist.gov/publications/fips/fips180-2/fips180-2.pdf. 1 Aug 2002
2. Antonopoulos, A.M.: Mastering Bitcoin: Unlocking Digital Cryptocurrencies, 1st edn. O'Reilly Media Inc., Sebastopol, CA, USA (2014)
3. Casado-Vara, R., Pierto, J., De la Pierta, F., Corchado, J.M.: International Workshop on IoT Approaches: For Distributed Computing, Communications and New Applications. IoTAS 2018
4. Wikipedia Contributors: https://en.wikipedia.org/wiki/Blockchain
5. Jabbari, A., Kaminsky, P.: Department of Industrial Engineering and Operations Research, University of California, Berkeley, Jan 2018

6. Wikipedia Contributor: https://en.wikipedia.org/wiki/Blockchain
7. NPTEL Course on Blockchain Architecture Design and Use Cases
8. Udemy Contributor: Blockchain A-Z-learn how to build your first Blockchain: Section-3: Blockchain Intuition, Immutable Ledger. https://www.udemy.com/build-your-blockchain-az/learn/v4/t/lecture/9657362?start=345
9. Udemy Contributor: Blockchain A-Z-learn how to build your first Blockchain: Section-3: Blockchain Intuition: Distributed P2P Network. https://www.udemy.com/build-your-blockchain-az/learn/v4/t/lecture/9657372?start=540
10. Udemy Contributor: Blockchain A-Z-learn how to build your first Blockchain
11. Udemy Contributor: Blockchain A-Z-learn how to build your first Blockchain: Section-3: Blockchain Intuition: Byzantine Fault Tolerance. https://www.udemy.com/build-your-blockchainaz/learn/v4/t/lecture/9657382?start=615
12. Udemy Contributor: Blockchain A-Z-learn how to build your first Blockchain: Section-3: Blockchain Intuition: Consensus Protocol. https://www.udemy.com/build-your-blockchain-az/learn/v4/t/lecture/9657384?start=0
13. Zheng, Z., Xie, S., Die, H.-N., Chen, X., Wang, H.: Blockchain challenges and opportunities: a survey. Int. J. Web Grid Serv **14**(4), 352 (2018)
14. Tutorial Point Contributor: Supply chain management. Available online https://www.tutorialspoint.com/supply_chain_management/supply_chain_management_quick_guide.htm
15. Verhoeven, P., Sinn, F., Herden, T.T.: Department of Logistics, Berlin Institute of Technology, 10623, Berlin, Germany, 11 Sept 2018
16. Lu blockchain—Internet of things supply chain traceability paper by Deloitte
17. Wikipedia Contributor: Waste management system. https://en.wikipedia.org/wiki/Waste_management
18. Health Library for Disasters: Solid waste management. http://helid.digicollection.org/en/d/Js2669e/7.7.4.html
19. Takyar, A.: CEO of LeewayHertz: Blockchain Technology and Business Book. Available online https://www.leewayhertz.com/blockchain-waste-management
20. WHO food safety statistics: Available online: https://www.who.int/news-room/fact-sheets/detail/food-safety
21. Blockchain in supply chain management system: Available online: https://www.bitdeal.net/blockchain-in-scm-supply-chain-management
22. What is supply chain management on blockchain: Available online: https://coinsutra.com/supply-chain-management-blockchain/
23. How blockchain disrupting supply chain industry. Available online: https://hackernoon.com/how-is-blockchain-disrupting-the-supply-chain-industryf3a1c599daef
24. A Deloitte Research Paper: When two chain combines supply chain meets blockchain
25. 3 ways blockchain will change the pharmaceutical supply chain. Available online: https://kinapsea.com/…/3_ways_blockchain_will_change_the_pharmaceutical_supply_chain
26. Nakasumi, M.: Information sharing for supply chain management based on blockchain technology: In: 2017 IEEE 19th Conference on Business Informatics. Faculty of Economics. Komazawa University, Japan

APDRChain: ANN Based Predictive Analysis of Diseases and Report Sharing Through Blockchain

Snehasis Bagchi, Mohuya Chakraborty and Arup Kumar Chattopadhyay

Abstract A huge amount of healthcare data (structured and unstructured) is currently available to medical specialists, indicating details of clinical symptoms. Each type of data provides information that must be properly analyzed for health-care diagnosis. To simplify the diagnostic process, avoid misdiagnosis as well as early detection, artificial intelligence (AI) that aims to mimic human cognitive functions may be employed. Current AI techniques that are used for structured data include machine learning methods, such as the classical support vector machine, artificial neural network, and the modern deep learning. Natural language processing is mainly used for unstructured data. In this paper, we have adopted artificial neural network by using adaptive learning algorithms to handle diverse types of cardiovascular clinical data and integrate them into categorized major cardiovascular disease outputs such as heart failure, aortic aneurysm, cardiomyopathy, cerebrovascular disease, etc. These outputs are then shared as reports to patients as well as doctors by an efficient report sharing scheme called APDRChain, which combines blockchain and structured peer-to-peer network techniques with clever cryptography to create a consensus mechanism. The evaluation results show that APDRChain can achieve higher efficiency and satisfy the security requirements in report sharing.

Keywords Blockchain · Cryptography · Artificial neural network · Healthcare data · Report sharing · Medical diagnosis

S. Bagchi · M. Chakraborty (✉) · A. K. Chattopadhyay
Institute of Engineering and Management, Kolkata 700091, India
e-mail: mohuyacb@iemcal.com

S. Bagchi
e-mail: snehasisbagchi@yahoo.com

A. K. Chattopadhyay
e-mail: arup.chattopadhyay@iemcal.com

© Springer Nature Singapore Pte Ltd. 2020
M. Chakraborty et al. (eds.), *Proceedings of International Ethical Hacking Conference 2019*, Advances in Intelligent Systems and Computing 1065, https://doi.org/10.1007/978-981-15-0361-0_8

1 Introduction

In today's world, Artificial Neural Networks (ANNs) are widely used in various applications of science and technology including computer science, communication engineering, mathematics, chemistry, physics, and biology. ANN is a computational model based on the structure and functions of biological neural networks. A neural network allows a machine to have cognitive functions to learn, deduce, and become accustomed to based on data it collects. As such, data flowing through the network affects its structure.

ANNs provide a dominant tool to help in predictive analysis of complex clinical data for diagnosis of wide range of diseases. Here, classification of diseases is done on the basis of the measured features to assign the patient to one of a small set of classified diseases.

Upon disease diagnosis, the doctor may further need to get the clinical history of the patient. Cloud-based healthcare data sharing schemes are in existence today. However, due to potential risk factors, patients do not want to transfer their confidential data to the cloud [1]. Blockchain can provide the same features much more efficiently without the need for trusting a third party. As per survey by the Global Blockchain Business Council (GBBC) reported by IBS Intelligence [2], effect of using blockchain technology in healthcare industry would be tremendous. There are numerous problems in healthcare that blockchain can help solve, such as data exchange limitations, supply chain issues, and patient data use hurdles in clinical research. Blockchain's structure provides a decentralized and secure way to exchange and track data to aid in addressing these problem. The advantages of blockchain for healthcare are as follows.

- Transparency: Data stored on the blockchain are transparent to approved users, creating a single source of truth.
- Trust: Data is linked through secure blocks that are distributed across multiple users, enabling trust between users who might not know each other.
- Disintermediation: Blockchain fulfills the role of existing intermediaries by creating an ecosystem of trust.
- Auditability: Data on the blockchain are difficult to change, creating a comprehensive audit trail.

In implementing blockchain for healthcare, the corresponding organizations should confirm as to who should have access to the ledger and what data will be accessible. Blockchain members can approve transactions and confirm identities and ownership with the requirement for third-party intermediaries. All related data may be shared with other members depending on their particular roles and access privileges. Blockchain networks often use encryption to ensure data security and privacy, which helps stop unauthorized access to transaction data and deter fraud.

This paper deals with predictive analysis of cardiovascular disease by using ANN and sharing of analysis report by a scheme called APDRChain that combines blockchain and structured peer-to-peer network techniques with clever cryptography to create a consensus mechanism.

The organization of the paper is as follows. After the introduction in Sect. 1, brief overviews of ANN and blockchain are presented in Sect. 2. Section 3 presents the ANN model for medical diagnosis of cardiovascular disease. Section 4 gives the overview and simulation result of ANN based Predictive Analysis and medical diagnosis report sharing blockchain model called APDRChain. In Sect. 5, we conclude the paper and present a glimpse of the future work.

2 Overview of ANN and Blockchain

2.1 Ann

Motivated by biological neural networks that make up animal brains, ANNs are used as computing systems to perform various computational tasks including complex mathematical problems, data classification, data clustering as well as regression. Learning mechanisms forms part of ANNs to perform tasks just like human brains by bearing in mind examples without getting programmed by any job specific rules. An ANN is based on a collection of interconnected artificial neurons, which replicate the neurons in a human brain. Each connection can transmit a signal from one artificial neuron to another just like the synapses in a human brain. On receiving the signal, an artificial neuron may do the internal processing of the signal and then pass it on to the other artificial neurons linked to it [3].

For implementing an ANN for a particular application, a real number is used as an input signal at a connection between artificial neurons, whereas the output of each artificial neuron is calculated based on some non-linear function of the sum of its inputs by the use of biases and weights well described mathematically according to computational model of McCulloch and Pitts4 as shown in Fig. 1 [4] and as given in Eq. (1) as per binary threshold unit. This calculates a weighted sum of its m input signals, s_i (where $i = 1, 2,..., m$) and generates an output of logic one if this sum is above a certain threshold "t," otherwise, an output of logic zero. Mathematically,

$$z = U\left\{ \sum_{i=1}^{n}(w_i * s_i) - t \right\} \tag{1}$$

where z is the output, $U\{.\}$ is a unit step function, and w_i is the synapse weight linked to the ith input.

As the process of learning continues, the weight adjusts itself by either enhancing or lessening the signal strength at a link. A threshold may be sought of such that the signal is sent to the next neuron if the collective signal crosses that threshold. Usually, ANNs have different layers consisting of a large number of neurons. These layers are responsible for carrying out different types of transformations on their inputs. Signals travel from the first layer (the input layer) to the last layer (the output layer), and possibly after moving back and forth the intermediate hidden layers many

Fig. 1 McCulloch–Pitts
model of a neuron

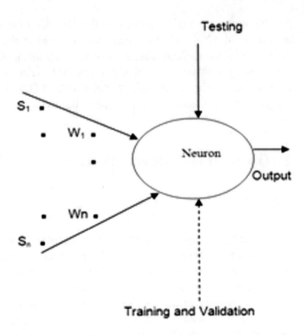

number of times. ANNs have found various usages in varied applications like medical diagnosis, machine translation, computer vision, playing board and video games, speech recognition, and spectrum sensing in cognitive radio.

2.2 Blockchain

A blockchain is an ever-increasing proof of records, called blocks, which are interconnected by using secured hash functions. Each block contains a cryptographic hash of the previous block, a timestamp, and transaction data. No alteration of data is possible in blockchain as per design methodology. It is an open, distributed ledger that can record transactions between two parties efficiently and in a verifiable and permanent way. Each of these blocks of data is protected and connected to one another other by the use of encryption techniques (i.e., chain). The blockchain network is decentralized in nature and due to its open, shared, and immutable nature; anyone and everyone may see the information in it. Whatever thing is developed on the blockchain is by its natural character clear, and all entities participating in it are responsible for their actions. Originally meant for the digital currency, Bitcoin, many other applications of blockchain are coming up in today's world [5].

3   ANN Model for Medical Diagnosis

Healthcare systems in today's world are transforming into a value oriented, patient-centric system of healthcare operation. As such, we visage new challenges involving the improvement of the configuration and supervision of healthcare delivery, viz. enhancing incorporation of methods used for patient-oriented continual disease supervision [6]. Complex decision making across different industries can be accomplished by the rapid use of Artificial Intelligence (AI). However, there is a lack of path on selecting suitable methods or technologies customized to the healthcare industry [7].

Patient care and operations management require the involvement and interprocess communication between various stakeholders, viz. doctors, middle managers, senior level officers to make decisions on a medical data including disease analysis, treatment, etc., and non-medical information including budget and resources.

In this paper, we have used ANN model for classification of four different types of cardiovascular diseases as shown in Fig. 2. ANNs require large training sets for estimation of numerous weights involved in calculation. We have used medical datasets of cardiovascular diseases from UCI Machine Learning Repository [8].

4   APDRChain

ANN Based Predictive Analysis of Diseases and Report Sharing Through Blockchain (APDRChain) is an ANN based medical disease diagnosis system that classifies four different types of cardiovascular diseases based on the datasets obtained from UCI Machine Learning Repository [8] and subsequently sharing the report to the concerned persons like doctor, patients, etc., securely by using blockchain technology.

4.1   ANN Model for Disease Classification

The ANN model for the classification of diseases as shown in Fig. 2 has been simulated in MATLAB 2019a [9]. It consists of one input layer, one hidden layer and one output layer. The input layer consists of 14 numbers of inputs to which 14 attributes of cardiovascular feature vectors are fed as shown in Table 1. The output layer consists of four numbers of outputs that are capable of detecting four different types of cardiovascular diseases, viz. (1) Heart failure, (2) Aortic aneurysm, (3) Cardiomyopathy, and (4) Cerebrovascular disease. These outputs show a logic 1 for the presence of the disease and logic 0 for its absence.

Figure 3 shows the flowchart of APDRChain. At the beginning of the process, cardiovascular dataset obtained from UCI Machine Learning Repository is acquired,

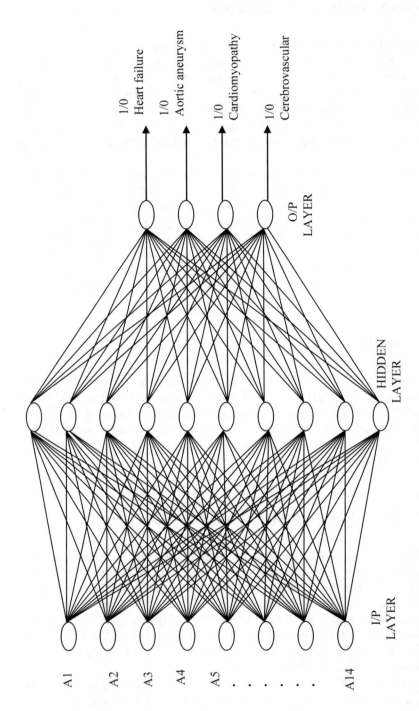

Fig. 2 ANN model for medical cardiovascular disease classification

Table 1 Cardiovascular disease attributes

Attribute name	Attribute number	Attribute details
A1	#3	age (years)
A2	#4	sex (male/female)
A3	#9	cp (type of chest pain)
A4	#10	trestbps (blood pressure during resting in mm Hg)
A5	#12	Chol (serum cholesterol in mg/dl)
A6	#16	fbs (fasting blood sugar >120 mg/dl)
A7	#19	Restecg (electrocardiographic results during resting)
A8	#32	thalach (maximum heart rate achieved)
A9	#38	exang (exercise-induced angina)
A10	#40	oldpeak (ST depression induced by exercise relative to rest)
A11	#41	slope (the slope of the peak exercise ST segment)
A12	#44	ca (number of major vessels)
A13	#51	thal
A14	#58	num (diagnosis of heart disease)

and data pre-processing is done to make the data compatible with MATLAB 2019a. At the next stage, training of ANN is done with the dataset. We have used 200,000 samples of data to train the ANN, and 250,000 samples were used as test data for validation. Next, the actual patient data was fed to the trained ANN model for matching and classification of diseases. If the result shows positive output, then report was generated and shared to doctor and/or patient by using blockchain model described later in Sect. 4.2.

The MATLAB Simulink ANN model for cardiovascular diseases classification which consists of 14 feature input vectors as per Table 1 and four outputs for classifying four different types of cardiovascular diseases with one hidden layer consisting of 12 neurons is given in Fig. 4. The number of hidden layers and number of neurons in the hidden layer has been fixed to make a trade-off between the accuracy and time complexity calculations during simulation process. Figure 5 shows the least mean squared error during training and validation. The confusion matrix obtained from the simulation results shows 92% accuracy with 3% false negatives and about 4% false positives. The result seemed to be encouraging.

4.2 Ethereum Blockchain Model for Report Sharing

Although in today date the most popular application of blockchain is in Bitcoin, however, blockchain has its enormous usage in building any non-financial decentralized application. Whenever medical data of patients are shared on a public domain,

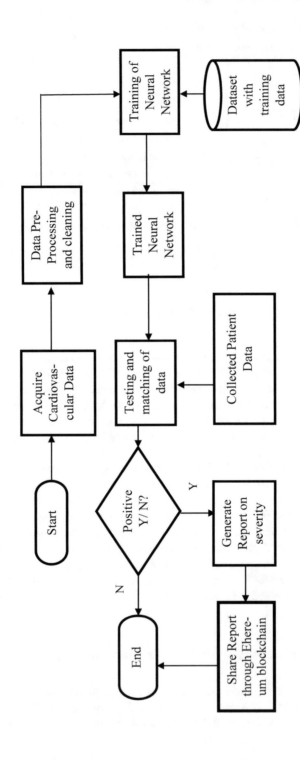

Fig. 3 Flowchart APDRChain

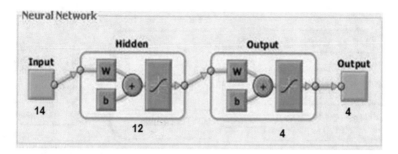

Fig. 4 MATLAB Simulink ANN model for cardiovascular disease classification

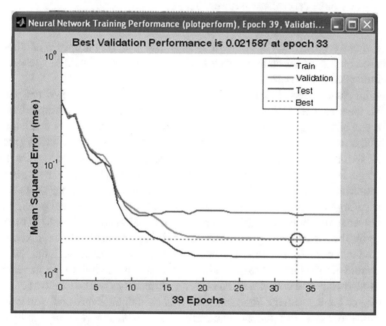

Fig. 5 Least mean squared error

there is a great possibility of these being misused by malevolent users causing huge financial loss as well as reputation damage to the stakeholders. The harms of distribution of health diagnosis reports are identified in [10]. Azaria et al. [11] proposed a decentralized record management system using Ethereum blockchain and named it as MedRec where patients would have an easy access to their medical information by ensuring verification, privacy, responsibility, etc. Xia et al. [12] gave the proposal for another data sharing architecture based on blockchain which uses the immutable and built-in autonomy attributes of blockchain and has used cryptographic keys to

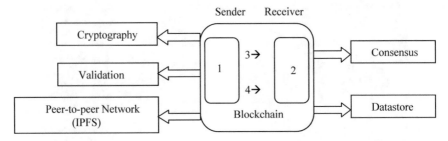

Legend: 1 → Document Storage; 2→ Smart Contract; 3 → Document Hash; 4 → Encryption Keys

Fig. 6 Blockchain model on Ethereum platform

ensure identity verification of different entities. Xia et al. [13] proposed another system based on blockchain that takes care of sharing of clinical data in a less-trusted situation called MeDShare.

The blockchain-based distributed immutable ledger is well suited for publishing. We have created Ethereum blockchain environment where medical diagnosis reports can be transferred securely without the need of any third-party mediator. The proposed concept consists of two key units: document storage and a smart contracts journal. We have used a private network called the InterPlanetary File System (IPFS) for storage. IPFS is a peer-to-peer distributed file system. The files can be uploaded by the stakeholders with an IPFS client into the IPFS and shared among the peers. We have used cryptographic technique to encrypt the secret key with the public key of the receiver so that only the receiver can decrypt it with his/her private key to ensure singing by the report issuing authority for auditing purpose and also token-based assets to the reports for patients. The encryption keys are sent through the Ethereum blockchain with the help of smart contracts that contain the link to the stored document, a secret key and sender's and receiver's addresses as well. The smart contracts journal used contains all the transactions as well as their attributes and is an interface for confidential use. The patients can further share their report to anyone by using a self-sovereign identity [14]. We propose to use the pairwise-pseudonymous identifiers, a separate Distributed Identifier (DID) for every relationship [14, 15]. The block diagram representation Ethereum blockchain is shown in Fig. 6 [16].

5 Conclusion

This paper has proposed an ANN based cardiovascular disease classification and detection model with subsequent report sharing by using blockchain to patients and doctors securely. Implementation results on MATLAB for the ANN based classification part showed 92% accuracy as given by the confusion matrix with about 3% false negatives and around 4% false positives. The Ethereum-based blockchain for model for report sharing proved to be efficient with respect to breaking the code. In future, we intend to make the system much more secure and efficient by the use of improved consensus algorithms. Moreover, real-time health monitoring of the patients is also to in the pipeline.

References

1. Shen, B., Guo, J., Yang, Y.: MedChain: efficient healthcare data sharing via blockchain. in Appl. Sci. **9**, 1207 (2019). https://doi.org/10.3390/app9061207
2. https://ibsintelligence.com/tag/global-blockchain-business-council/
3. Mishra, M., Srivastava, M.: A view of artificial neural network. In: Proceeding 2014 International Conference on Advances in Engineering & Technology Research (ICAETR-2014). https://doi.org/10.1109/icaetr.2014.7012785. IEEE Xplore 19 January 2015
4. Jain, A.K., Mao, J., Mohiuddin, K.M.: Artificial neural network: a tutorial. IEEE Comput., **29**, 31–44 (1996)
5. Zheng, Z., Xie, S., Dai, H.: Chen, X.: An overview of blockchain technology: architecture, consensus, and future trends. In: Proceeding 2017 IEEE International Congress on Big Data (Big-Data Congress). https://doi.org/10.1109/bigdatacongress.2017.85. IEEE Xplore 11 September 2017
6. Kuziemsky C.: Decision-making in healthcare as a complex adaptive system. Healthc Manag. Forum. **29**(1):4–7. (2016). https://doi.org/10.1177/0840470415614842. [PubMed] [Cross-Ref] [Google Scholar
7. Deloitte.: Global health care outlook: The evolution of smart health care (2018)
8. https://archive.ics.uci.edu/ml/datasets/heart+Disease
9. https://www.mathworks.com/products/matlab/whatsnew.html
10. da Conceicao, A.F., da Silva, F.S.C., Rocha, V., Locoro, A., Barguil, J.M.M.: Electronic health records using blockchain technology (2014)
11. Azaria, A., Ekblaw, A., Vieira, T., Lippman, A.: MedRec: using blockchain for medical data access and permission management. In: Proceeding 2nd International Conference on Open and Big Data, Vienna, Austria (2016)
12. Xia, Q., Sifah, E.B., Smahi, A. Amofa, S., Zhang, X.: BBDS: Blockchain-based data sharing for electronic medical records in cloud environments. Information, **8**(2) (2017)
13. Xia, Q., Sifah E.B., ASAMOAH, K.O., Gao, J., Du, X., Guizani, M.: MeDShare: trust-less medical data sharing among cloud service providers via Blockchain. IEEE Access, 5, 14757–14767 (2017)
14. Lim, S.Y., Fotsing, P.T., Almasri, A., Musa, O., Kiah, L.M., Ang, T.F., Ismail, R.: Blockchain technology the identity management and authentication service disruptor: A Survey, **8**(4), 1735–1745
15. Andrew Tobin, D.R.: The Inevitable rise of self-sovereign identity (White paper). 2017: Sovrin Foundation
16. https://www.ethereum.org/

Modeling and Simulation

Study on S-box Properties of Convolution Coder

S. RoyChatterjee, K. Sur and M. Chakraborty

Abstract The substitution box (S-box) is an essential component in cryptography. The desirable cryptographic properties of the S-box are bijective, non-linearity, strict avalanche, bit independence, and resiliency. The analyses of these properties determine the quality of S-box. This paper provides mathematical analysis of cryptographic properties of the convolution coder. Simulation results show that it satisfies most of the properties and robustness increases with the number of iteration.

Keywords Cryptography · S-box · Convolution code

1 Introduction

Cryptography plays a key role in information security. Among the different part of cryptography, substitution box plays a pivotal role in case of encryption. There are different techniques or algorithms used in substitution box to make a strong encryption [1, 2]. In [3], the authors utilize a chaotic system to generate S-box to make it more sensitive in randomness of the primary condition. In [4], the authors discussed the analogy between the design of S-box and traveling salesman problem. There the entities in the sequence of S-box are analog with the cities. They proposed genetic algorithm for designing of S-box. In [5], the authors proposed design of S-box with Boolean function considering each component of the input vector to Boolean function that is independent and identically distributed Bernoulli variates with the parameter. In [3], the look-up tables based static S-box has been designed. However, the results show that it may not suitable for high-speed applications. In [6], the authors

S. RoyChatterjee (✉)
Netaji Subhash Engineering College, Kolkata, India
e-mail: rcswagata@gmail.com

K. Sur · M. Chakraborty
Institute of Engineering & Management, Kolkata, India
e-mail: kajari.sur@iemcal.com

M. Chakraborty
e-mail: mohuyacb@iemcal.com

© Springer Nature Singapore Pte Ltd. 2020
M. Chakraborty et al. (eds.), *Proceedings of International Ethical Hacking Conference 2019*, Advances in Intelligent Systems and Computing 1065, https://doi.org/10.1007/978-981-15-0361-0_9

designed the S-box circuit utilizing truth table representing the value of output with the changes of input values and made the circuit with computational logic units. Other methodology includes multiplicative reversal in Galois Field utilizing composite field [7]. However, it requires high delay, and it has high equipment complexities.

Though there are different algorithms for S-box, all techniques are supposed to be abided by the desirable properties of S-box which include bijective property, non-linearity property, strict avalanche criterion, bit independence, and resiliency property [8, 9].

On the other hand, there is well-known encoder called convolution coder [10]. It is widely accepted for the error detection in wireless transmission. It produces new output bit with the variation of the input bit pattern. At the same time, it utilizes one to many mapping and easy to implement in hardware. Because of these natures of convolution coding technique, it may be suitable for designing non-linear S-box This paper analyzes cryptographic properties of convolutional coder to testify its efficiency in designing S-box.

After introduction, desirable properties of S-box are discussed in Sect. 2. In Sect. 3, convolutional coder is represented by Boolean function, and analytical analysis is performed to testify properties of the S-box. Section 4 provides simulation results followed by the conclusion in Sect. 5.

2   Features of S-box

Different methods are employed to measure the strength of the S-box. In paper [11], the authors established selection process which helps to evaluate the quality of the S-box. The guideline is provided by National Security Agency [12] to assess the cryptographic properties of S-box. They are bijective property, the non-linearity, the strict avalanche criterion (SAC), the bits independence criterion (BIC), and the resiliency property.

Bijective Property:. Mathematically, a bijective function $f: X_n \rightarrow Y_n$ is injective (one-to-one) and surjective (onto mapping) of a set X_n to a set Y_n.

Non-Linearity Property: Non-linearity for a binary stream is given by non-linearity value, and Nf must be less than or equal $2^{n-1} - 2^{\frac{n}{2}-1}$, where n is the bit length and

$$N_f = 2^{n-1} - \frac{1}{2} \max \ W_f(u), \text{ where } u \epsilon f_2^n, \ W_f(u)$$

$$= \sum_{x \epsilon f_2^n} f(x)(-1)^{\langle x,u \rangle} \text{ is a Walsh Spectrum of } f \ \& \ c \in \{0, 1\}^n.$$

Strict Avalanche Criterion (SAC): It reflects the probability of the change of output bit with the changes of the input bit. Non-linear functions satisfy higher order SAC [13].

Bit Independence Criterion (BIC): It is another parameter used to test the security of S-boxes in term of robustness against various attacks.

Resiliency Property: Resiliency property suggests that Boolean functions are balanced and correlation immune, i.e., their Walsh Spectrums satisfy $W_f(u) = 0$, where $1 \leq wt(u) \leq m$.

3 Analytical Description of Convolution Coder

Convolutional coder is a memory-based circuit, and it utilizes Mod 2 addition operation between present and previous value of input bit stream to generate output bit pattern. Figure 1 shows that output bits (o'_n) and (o_n) generate from the input bit stream (i_n). As a result, the substitution operation produces double number of bit than that of the input bit stream.

The Boolean function of the Convolution Coder is

$$f_H(x) = \sum_{n=0}^{n=m-3} \left\{ (H_n \oplus H_{n+1} \oplus H_{n+2})x^{2(m-1-n)} \right.$$
$$\left. + (H_n \oplus H_{n+2})x^{2(m-1-n)+1} \right\}$$
$$+ (H_{m-2} \oplus H_{m-1} \oplus b_1)x^3 + (H_{m-2} \oplus b_1)x^2$$
$$+ (H_{m-1} \oplus b_1 \oplus b_0)x + H_{m-1} \oplus b_0$$

The even function of the Convolution Coder is

$$f_{eH}(x) = \sum_{n=0}^{n=m-3} \left\{ (H_n \oplus H_{n+1} \oplus H_{n+2})x^{m-1-n} \right\}$$
$$+ (H_{m-2} \oplus H_{m-1} \oplus b_1)x + H_{m-1} \oplus b_1 \oplus b_0$$

The odd function of the Convolution Coder is

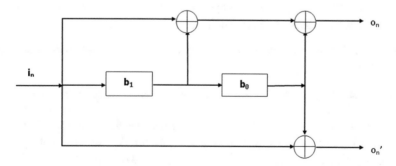

Fig. 1 (2, 1, 2) convolutional coder

$$f_{\text{oH}}(x) = \sum_{n=0}^{n=m-3} \left\{ (H_n \oplus H_{n+2}) x^{m-1-n} \right\}$$

$$+ (H_{m-2} \oplus b_1)x + H_{m-1} \oplus b_0$$

where m is the number of input bits b_0 and b_1 which are the initials of the registers of the coder, and $\{A_{m-1}, A_{m-2} \ldots A_1, A_0\}$ are the coefficients of the input binary string. The generated bit stream is subdivided into odd bit stream and even bit stream.

Theorem 1 *A Boolean function satisfies bijective property if both injective property, which is one-to-one correspondence between domain and co-domain, and surjective property, which demonstrates about same co-domain for different domains, are satisfied.*

Proof Let two different input strings B and C. $B = \{B_0, B_1, \ldots, B_{m-2}, B_{m-1}\}$, $Q = \{C_0, C_1, \ldots, C_{m-2}, C_{m-1}\}$ are as domains. Since the degrees of coefficients are 1, different combinations of strings produce its own respective expressions until they are equal as in any Boolean expression, odd degree of any term in the expression yields different results for different domains, and even degree of any term in the expression may yield same results for different domains. Hence, both the functions satisfy injective property.

$$f_{\text{eH}}(x) = \begin{cases} H_0 \oplus b_0, \forall x \in I \ \& \ x \quad \mod 2 = 1 \\ H_{m-1} \oplus b_1 \oplus b_0, \forall x \in I \ \& \ x \quad \mod 2 = 0 \end{cases}$$

$$f_{\text{oH}}(x) = \begin{cases} H_{m-1} \oplus H_1 \oplus b_1 \oplus b_0, \forall x \in I \ \& \ x \quad \mod 2 = 1 \\ H_{m-1} \oplus b_0, \forall x \in I \ \& \ x \quad \mod 2 = 0 \end{cases}$$

For different odd domains of x, the co-domains or output for even function : $f_{\text{eH}}(1) = H_0 \oplus b_0$ & for odd function, $f_{\text{oH}}(1) = H_{m-1} \oplus H_1 \oplus b_1 \oplus b_0$, for even different domains of x, for even function: $f_{\text{eH}}(0) = H_{m-1} \oplus b_1 \oplus b_0$ & for odd function $f_{\text{oA}}(0) = H_{m-1} \oplus b_0$ are similar, thus satisfying surjective property. Hence, both even function and odd function of the expansion box satisfy bijective property as both injective property and surjective are satisfied.

Theorem 2 *For satisfaction of non-linearity property, non-linearity N_f must lesser than or equal to $2^{n-1} - 2^{\frac{n}{2}-1}$.*

Proof Considering the even function and applying the conditions of non-linearity property on it, $N_f \leq 2^{n-1} - 2^{\frac{n}{2}-1}$, where $N_f = 2^{n-1} - \frac{1}{2} \max W_f(u)$, where $u \in f_2^n$, $W_f(u) = \sum_{x \in f_2^n} f(x)(-1) \langle x, u \rangle$ is a Walsh Spectrum of f & $c \in \{0, 1\}^n$ with respect of which Walsh Spectrum of f is to obtain. We get, $N_{f_{\text{eH}}} = 2^{n-1} - \frac{1}{2} \max W_{f_{\text{eH}}}(u)$, where Walsh Spectrum of f_{eH} $W_{f_{\text{eH}}}(u) = \sum_{x \in f_2^n} f_{\text{eH}}(x)(-1)^{\langle x, u \rangle}$. As the field has various combinations of 0's and 1's, the expression must contain even number of bits, and coefficients must be 1, $W_{f_{\text{eH}}}(u) = 2^{m/2}$.

$$\because N_f = 2^{m-1} - \frac{1}{2}\max W_f(u) = 2^{m-1} - 2^{\frac{m}{2}-1}, \text{ where } m \text{ is the number of bits}$$

So, maximum value of N_f is $2^{m-1} - 2^{\frac{m}{2}-1}$. Again for odd function also, the maximum value of N_f is $2^{m-1} - 2^{\frac{m}{2}-1}$. For $m = 2$, maximum $N_f = 2^{2-1} - 2^{\frac{2}{2}-1} = 1$. Hence, for both the even and odd functions, non-linearity property is satisfied. For the non-linearity property, the relation of the decimal values for the corresponding output binary streams with its non-linearity values has been depicted.

Theorem 3 *Any Boolean function satisfies strict avalanche criterion or SAC if $f(x + \alpha) - f(x) = 0$, i.e., f(x) is a balanced function of any order n and $\alpha \in GF(2^n)$ for that value of n and its wt $(\alpha) = 1$.*

Proof For the even function for achieving the conditions satisfying SAC, $f_{\text{eH}}(x + \alpha) = f_{\text{eH}}(x)$ So, the satisfied condition is $\sum_{n=0}^{n=m-3} H_n \oplus H_{n+1} \oplus H_{n+2} + H_{m-2} \oplus H_{m-1} \oplus b_1 = 0$. For the odd function for achieving the conditions satisfying SAC, $f_{\text{oH}}(x + \alpha) = f_{\text{oH}}(x)$ So, the satisfied condition is $\sum_{n=0}^{n=m-3} H_n \oplus H_{n+2} + H_{m-2} \oplus b_1 = 0$.

Theorem 4 *Bit independence criterion or BIC is satisfied if $f_c(x + \alpha) - f_c(x) = 0$, where f(x) is a Boolean function of any order n and satisfies SAC, $\alpha \in GF(2^n)$ for that value of n and its wt $(\alpha) = 1$, and c is any combination where $c \in \{0, 1\}^n$ of any order n and wt$(c) \geq 1$.*

Proof Let us consider a string R satisfying SAC. For even function for achieving the conditions satisfying BIC, $f_{\text{eR}}(x + \alpha) = f_{\text{eR}}(x)$ So, the satisfied condition is $\sum_{n=0}^{n=m-3} R_n \oplus R_{n+1} \oplus R_{n+2} + R_{m-2} \oplus R_{m-1} \oplus b_1 = 0$. For odd function for achieving the conditions satisfying BIC, $f_{\text{oR}}(x + \alpha) = f_{\text{oR}}(x)$ So, the satisfied condition is $\sum_{n=0}^{n=m-3} R_n \oplus R_{n+2} + R_{m-2} \oplus b_1 = 0$. The bit independence criterion is satisfied as both the even and odd functions are balanced function.

Theorem 5 *The odd and even function of order m satisfy resiliency property if they are balanced & correlation immune, i.e.,. their Walsh Spectrums satisfy $W_f(u) = 0$, where $1 \leq wt(u) \leq m$. Here, the function is balanced if $W_f(0) = 0$.*

Proof Considering the even function and it's Walsh Transform is $W_{f_{\text{eH}}}(u) = \sum_{x \in f_2^n} f_{\text{eH}}(x)(-1)^{\langle x, u \rangle}$. Hence, for the function to be balanced, $\sum_{n=0}^{n=m-3} H_n \oplus H_{n+1} \oplus H_{n+2} + H_{m-2} \oplus H_{m-1} \oplus b_1 = 0$. For correlation immune,

$$W_{f_{\text{eH}}}(u) = (1 + 2 + \cdots + m - 1) \sum_{n=0}^{n=m-3} H_n \oplus H_{n+1} \oplus H_{n+2}$$
$$+ m(H_{m-2} \oplus H_{m-1} \oplus b_1)$$
$$- \left((1 + 2 + \cdots + m - 1) \sum_{n=0}^{n=m-3} H_n \oplus H_{n+1} \oplus H_{n+2} \right.$$
$$\left. + m(H_{m-2} \oplus H_{m-1} \oplus b_1) \right) = 0$$

Considering the odd function and it's Walsh Transform is $W_{f_{oH}}(u) = \sum_{x \in f_2^n} f_{oH}(x)(-1)^{\langle x,u \rangle}$. Hence, for the function to be balanced, $\sum_{n=0}^{n=m-3} H_n \oplus H_{n+2} + H_{m-2} \oplus b_1 = 0$. For correlation immune,

$$W_{f_{eH}}(u) = (1+2+\cdots+m-1) \sum_{n=0}^{n=m-3} H_n \oplus H_{n+2} + m(H_{m-2} \oplus b_1)$$

$$- \left((1+2+\cdots+m-1) \sum_{n=0}^{n=m-3} H_n \oplus H_{n+2} + m(H_{m-2} \oplus b_1) \right) = 0$$

Since for a given condition for balance property & proof of correlation immune, resiliency is proved for both even & odd functions. Since the resiliency property suggests zeroes of the Walsh Transform of the binary stream, the number of zeroes of the Walsh Transform of the output bit stream of the decimal values is determined.

4   Simulation and Result Analysis

We consider two scenarios where the initial values of the registers as 1, 0 and in second case 0, 1. Figure 1 shows the behavior of bijective property for all the decimal numbers corresponding to output binary streams with its frequency of occurrences. Here, in Fig. 2, at 1st Iteration, very few of the decimal numbers have its frequency of occurrences as 1. So, in all the stages, the system is not satisfying bijective property. In Fig. 3, here also like the previous one, in all the four stages, only one value is having frequency of occurrence as 1, and the rest are having it different from 1. So, in all the stages, the system is not satisfying bijective property. Here, the behavior of

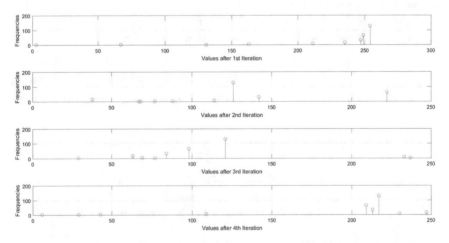

Fig. 2 Behavior of the bijective property with initial values of 1st and 2nd Registers as 1 and 0

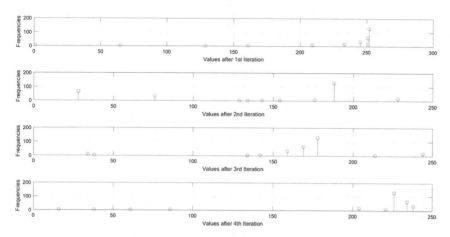

Fig. 3 Behavior of the bijective property with initial values of 1st and 2nd Registers as 1 and 0

bijective property is independent of initial values of 1st and 2nd Registers as the behavior for the above cases produces the same result, though the position the value having frequency of occurrence as 1 gets shifted with the initial values. In Fig. 4, it is seen that in the case of registers with initial value 1 and 0, respectively, the behavior of decimal values with its non-linearity values for 8-bit length. Here, the maximum non-linearity value is 120. In this case, only 4th iteration is preferred than the previous three as it produces the number of values with non-linearity much greater than the previous three. In Fig. 5, it is seen that in the case of registers with initial value 1 and 0, respectively, in the 1st iteration, the behavior of the Hamming weights with the frequency of the occurrences of Hamming weights shows monotonic, in the 2nd

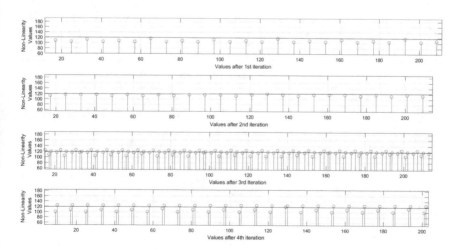

Fig. 4 The relation of the values with non-linearity values with the initial values of 1st and 2nd Registers as 1 and 0. N.B: The dark line represents the maximum non-linearity value

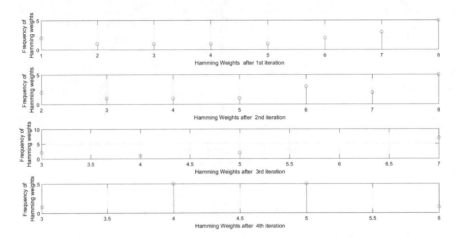

Fig. 5 Behavior of the Hamming weights with the frequency of occurrences of Hamming weights with initial values of 1st and 2nd Registers as 1 and 0

iteration, the behavior shifts from monotonic, and in the 3rd and 4th iterations, the behavior shows non-monotonic. 1st iteration is preferred than the next three as the behavior is monotonic. In Fig. 6, it is seen that in the case of registers with initial value 0 and 1, respectively, all the iteration shows monotonic. Also, 1st condition with initial values of 1st and 2nd Registers as 1 and 0 and key elements as 11111111 are preferred as it, only, produces the behavior to be monotonic. In Fig. 7, it is seen that in the case of registers with initial value 1 and 0, respectively, 1st iteration is preferred than the three iterations as it produces more resilient values than others. In Fig. 8, it is seen that in the case of registers with initial value 0 and 1, respectively,

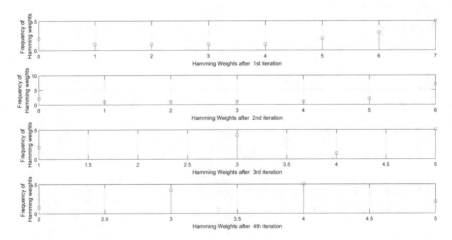

Fig. 6 Behavior of the Hamming weights with the frequency of occurrences of Hamming weights with initial values of 1st and 2nd Registers as 0 and 1

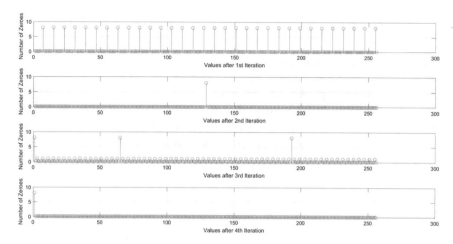

Fig. 7 Decimal values with number of zeroes with initial values of 1st and 2nd Registers as 1 and 0

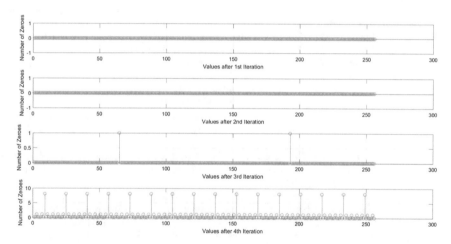

Fig. 8 Decimal values with number of zeroes with initial values of 1st and 2nd Registers as 0 and 1

4th iteration is preferred than the other three iterations as it produces more resilient values than the three. For circuit implementation, 1st condition with initial values of 1st and 2nd Registers as 1 and 0 is preferred as 1st condition requires only one box as 1st iteration is more resilient than the three, and 2nd condition requires four boxes as the 4th iteration is more resilient than the three.

5 Conclusion

This paper provides comprehensive mathematical analysis of cryptographic properties of convolutional coder. It is represented by the Boolean odd and even function, and mathematical study justifies that it satisfies most of the properties. Simulations with different initial conditions are performed to examine the nature of the changes of the properties with number of iterations. Results show that robustness increases with the number of iteration. So, this may be effective in designing dynamic S-box for encryption algorithms.

References

1. Hussain, I., Shah, T., Mahmoodm, H., Gondal, M.A.: A projective general linear group based algorithm for the construction of substitution box for block ci-phers. Neural Comput. Appl. **22**(6), 1085–1093 (2013)
2. Khan, M., Shah, T., Gondal, M.A.: An efficient technique for the construction of substitution box with chaotic partial differential equation. Nonlinear Dyn., **73**(3), 1795–1801 (2013)
3. Garg, S., Upadhyay, D.: S-box design approaches: critical analysis and future directions. Int. J. Adv. Res. Comput. Sci. Electron. Eng. (IJARCSEE) **2**(4) (2013)
4. Wang, Y., Wong, K.-W., Li, C., Li, Y.: A novel method to design S-box based on chaotic map and genetic algorithm. Phys. Lett. A **376**, 827–833(2012)
5. Gangopadhyay, S., Gangopadhyay, A.K., Pollatos, S., et al.: Cryptographic Boolean func-tions with biased inputs Cryptogr. Commun. **9**(2), 301–314 (2017)
6. Rachh, R.R., Ananda Mohan, P.V.: Implementation of AES S-boxes using combinational logic. In: IEEE International Symposium on Circuits and Systems, 2008. ISCAS 2008, pp. 3294–3297 (2008)
7. Nalini, C., Anandmohan, P.V., Poomaiah, D.V., Kulkarni, V.D.: Compact designs of subbytes and mixcolumn for AES. In: Advance Computing Conference, 2009. IACC 2009. IEEE International, pp. 1241–1247 (2009)
8. Mohamed, K., Mohammed Pauzi, M.N., Hj Mohd Ali, F.H., Ariffin, S., Nik Zulkipli, N.H.: Study of S-box properties in block cipher. In: 2014 International Conference on Computer, Communications, and Control Technology (I4CT), pp. 362–366, Langkawi (2014)
9. Dey, S., Ghosh, R.: A review of cryptographic properties of S-boxes with generation and analysis of crypto secure S-boxes. PeerJ Preprints **6**, e26452v1 (2018)
10. Baby, R.A.: Convolution coding and applications: a performance analysis under AWGN channel. In: 2015 International Conference on Communication Networks (ICCN), pp. 84–88, Gwalior (2015)
11. Jakimoski, G., Kocarev, L.: Chaos and cryptography: block encryption ciphers based on chaotic maps. In: IEEE Transactions on Circuits and Systems I: Fundamental Theory and Applica-tions, vol. 48, no. 2, pp. 163–169 (2001)
12. Branstad, D.K., Gait, J., Katzke, S.: Report of the workshop on cryptography in sup-port of computer security. Tech. Rep. NBSIR 77–1291, National Bureau of Standards, Sept. (1976)
13. Webster, A.F., Tavares, S.E.: On the design of S-boxes. In: Williams H.C. (eds.), Advances in Cryptology—CRYPTO '85 Proceedings. CRYPTO 1985. Lecture Notes in Computer Science, vol. 218. Springer, Berlin, Heidelberg (1986)

Smart Healthcare Assistance Toward On-road Medical Emergency

Debajyoti Basu, Sukanya Mukherjee, Anupam Bhattacharyya,
Swapnaneel Dey and Ajanta Das

Abstract The advancement of mobile computing as a technology together with the advancement of network communications has led to the foundation of exhaustive usage of mobile devices and corresponding mobile apps embedded in them. These apps provide value-added services in different domains like transport, entertainment, education, business, economics, and a variety of subjects. The current paper focuses on the healthcare domain and attempts to enlist a few basic functionalities and transform them as a user-friendly app for availing medical help and facilities on the move. Some apps provide chat communication facilities while others help in search location attempts. The currently proposed app, MedTravel, includes novelties like integrating the channel of communication between healthcare professionals and patients/users, searching for medical facilities like hospitals, medicines, and doctors besides providing opportunities for medical practitioners, pharmacies, and hospitals to register. Besides providing a comparative study with other medical app providers, it enhances the scope and features and formulates into a robust app with more opportunities of integrating with other facilities besides increasing existing features and functionalities.

D. Basu
Department of Computer Science & Engineering, Birla Institute of Technology, Mesra,
Kolkata Extension Centre, Kolkata, India
e-mail: devjyotibasu@yahoo.co.in

S. Mukherjee · A. Bhattacharyya (✉) · S. Dey
Department of Computer Science & Engineering, University of Engineering & Management,
Kolkata, India
e-mail: anupam.bhattacharyya.uemkcse16@gmail.com

S. Mukherjee
e-mail: m.sukanya1997@gmail.com

S. Dey
e-mail: swapnaneeld13@gmail.com

A. Das
Amity Institute of Information Technology, Amity University, Kolkata, India
e-mail: cse.dr.ajantadas@gmail.com

© Springer Nature Singapore Pte Ltd. 2020   129
M. Chakraborty et al. (eds.), *Proceedings of International
Ethical Hacking Conference 2019*, Advances in Intelligent
Systems and Computing 1065, https://doi.org/10.1007/978-981-15-0361-0_10

Keywords Healthcare · Medical emergency · MedTravel

1 Introduction

Efficient clinical communication and ease of access to medical facilities are becoming increasingly important in the healthcare domain for mobile apps. Some mobile apps connect patients to providers; some provide access to medical health records to patients and doctors, while others provide performance monitoring mechanisms. Some apps even help if providing diagnostic help to medical practitioners. However, one of the challenges which the user and patient community still suffer from is lack of proper information on medical facilities while on the move. The source of information has to be authentic and should be provided at the time of necessity and emergency. The current research paper attempts to identify such areas of requirement which patients and users need very frequently and convert them into a tangible solution in the form of a mobile medical app.

The objective of this paper is to present the functionalities of the proposed medical app (*MedTravel*) consisting of three modules: doctors, hospitals, and medical shops, as mentioned in the following:

i.   The app is integrated with Google Maps and is able to search and locate nearby medical facilities for the user.
ii.   The user will have an opportunity to access this information, execute subsequent searches to avail this information.
iii. The healthcare service providers and medical practitioners will have the opportunity to register themselves to provide their services to the end users.

Organization of the paper is as follows: Related work is presented in Sect. 2. Section 3 presents importance of the medical app and its novelties and benefits. Evaluation of the developed app is explained in Sects. 4 and 5 concludes the paper.

2 Related Work

This section presents a study of a few existing research papers based on medical apps. Guanling Chen et al. in their paper MPCS: Mobile Phone-Based Patient Compliance System for Chronic Illness Care, discuss a mobile phone-based patient compliance system [1].

They achieve to integrate social behavior theories to ensure that patients suffering from chronic illness for example diabetes follow compliance norms to take care and of themselves and ensure proper monitoring of their health.

Kaushal Modi and Radha Baran Mohanty from Infosys, in their white paper on mHealth: Challenges, benefits, and keys to successful implementation, discuss various mHealth apps and providers, challenges and the scope of implementing mHealth [2]. The paper discusses the keys to successful implementation: user adoption of apps, measurement of user satisfaction, access of app from various devices, and identification of user needs.

On A Provably-Secure-Cross-Domain Handshake Scheme with Symptoms-Matching for Mobile Healthcare Social network that allows greater security scheme to register in two different healthcare centres to execute cross domain handshake through symptoms matching [3].

Darell M West in his paper on Improving Health Care through Mobile Medical Devices and Sensors a part of their Mobile Economy project in Center for Technology Innovation at Brookings discusses innovations in Mobile Healthcare. Difficulties in this sector are physical distance between doctors and patients and lack of health care equipment and infrastructure at the right time [4].

In their paper "Provisioning of Medical Analysis in Cloud," Ajanta Das et al. propose a mobile app integrated with Big Data and uses analytics as a service to provide users with required medical information while on the move [5].

In their paper "Big Data computing and clouds: Trends and future directions" Marcos D. Assuncao et al. discuss approaches and environments for carrying out analytics on clouds for Big Data apps and how businesses can benefit from usage of Big Data [6]. Wei-Tek Tsai, Xin Sun, JanakaBalasooriyain in their paper discuss the Service-Oriented Cloud Computing architecture to overcome problems of existing cloud architectures [7].

Mobile technologies provide the opportunity to connect patients with their doctors, medical shops and enable timely health monitoring which suggests improved patient engagement and better health outcomes. The proposed solution attempts to bridge this gap as a healthcare app but as novelty, it will integrate both users and medical practitioners and provide search facilities for medical logistics. The proposed app provides services along with registration opportunities to augment services further to both the static and mobile users. It means if the users are far away of his or her own destination, still they get the medical support through this app. More specifically, it will search for medicines in an emergency and check their availability in nearby shops. Also, it readily proposes to display the numbers of beds available in a specific hospital at a point of time. The app enables quick search of medical provisions and also gives a snapshot hospitals, doctors, and pharmacies around the user in any geographic location, which the user/patient can use to access the information and reach out for medical help immediately.

3 Proposed Mobile Medical Apps

With increased network connectivity and development of robust hardware for mobile devices, the growing importance and benefits of mobile app in all domains continue to increase. Healthcare being one such domain is assuming immense significance which continues to reap benefits from this technological advancement.

3.1 Proposal Study

A study was conducted to understand the scope of the app and the requirements the app would cater to. The requirement analysis conducted is based on the following broad proposals promulgated. The app, *MedTravel*, should have three broad modules, doctors, medical shops, and hospitals. It is required that the following requirements are met by the app

1. The doctor module should provide opportunities for new doctors to register.
2. The users should be able to search for doctors in emergency in nearby locations.
3. The medical shop module should provide users to search for medical shops and also to search for availability of specific medicines.
4. The hospital module should provide users to search for hospitals and also check for the availability of bed in the respective hospital.

The search functionality is proposed to be on distance and the results will be returned in ascending order of proximity to the user's location.

3.2 Architectural Analysis

A. On the basis of requirement study, an architecture of this proposed app, *Med-Travel,* is presented in Fig. 1.

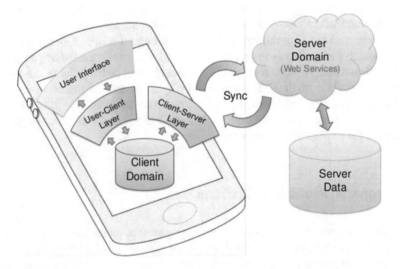

Fig. 1 Architecture of healthcare mobile app

The basic components are outlined here along with examples of their interactions. The app has been conceptualized on the basic client–server architecture of mobile apps.

User Interface → this forms the Graphical User Interface (GUI) of the app and is the main window of communication between all consumers of the app enabling users to get access to the core functionalities of the app.

User Client → this is the App Control Logic Layer. This layer forms the control logic and regulates the flow of the app; for example, if the user clicks on the doctor module, the control logic accepts the user input and accordingly displays the list of doctors.

Client-Server Layer → this is the App Networking Layer. It is through this layer that we interact with standard Google API's and along with other third-party API's. An app program interface (API) is a set of standards with which technical interfacing is possible to build software apps. To enable location-based search of doctors, hospitals, and healthcare facilities, we have integrated our app with Google Map APIs.

Client Domain → the technical components which are hosted in the users' handset defines the boundary of the Client Domain.

In the context of our app, the Persistence Data Storage/App Data Storage stores local data in the users mobile and constitutes the Client Domain. This can be referred to as local data storage which comes in play when there are connectivity issues. This is local persistence storage and is synced up with the main server database at regular intervals.

Server Domain → The Server Domain of the architecture constitutes the components hosted in the Server and is in principle in accordance with the Client–Server Architecture. In the Server Domain, in our app, we propose to use the Cloud API Server. The Cloud API Server—this is the main server where the app is proposed to be hosted. When any user downloads the app, the download takes place from here.

Server Data → the data persistence layer at the Server end is referred to as the Server Data layer. Here this constitutes the Cloud Database which is SQL Lite in our case.This interacts with Cloud Server and serves as the main Server persistence store of the app.

In order to illustrate how these architectural components interact, we take the example of the new doctor registration process.

(1) The new data is entered using the User Interface by the doctor.
(2) The User–Client Layer persists this data in the Client Domain.
(3) As soon as network connectivity is restored, the Client–Server Layer synchronizes this data with the Server Domain.
(4) The Server Domain finally pushes this data into the Server Data (SQL Lite).
(5) This data then becomes available and is cached back to the Client Domain from the Server Domain via the synchronization process, so the next time a user accesses the app via the User Interface, the User Client Layer pulls this data from Client Domain and makes the newly registered doctor's data available.

3.3 Functionalities of MedTravel

The app, *MedTravel,* has the following major functionalities:

1. It allows the user to conduct specialized search for medicine, doctors, and hospitals.
2. It allows doctors, hospitals, and medical shops to register to make their services available.
3. It utilizes third-party Google APIs to provide search and locate functionalities of registered medical facilities and practitioners.

Figure 2 explains the Use Case Diagram focusing on the functionalities and the principal actors of the app.

The Use Case Diagram enlists the following actors:

1. User: These are the general users of the app and includ patients and any person downloading and using the app.
2. Doctors: The doctors can register in the app and provide healthcare services.
3. Medical shops: The pharmacies are enlisted as actors and the app enables them to enlist via the registration process and also to enlist the availability of medicines
4. Hospitals: The hospitals are actors and post-registration are proposed to be displayed when the user searches for them.

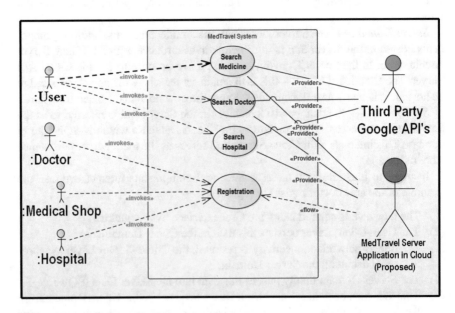

Fig. 2 Use case diagram of healthcare app

5. Third-party Google APIs: These API's are actors that augment the core functionalities of the map by providing location facilities.
6. *MedTravel* Server: The hosting facilities of the app are provided by the app server hosted in the cloud.

These actors interact with the User Client and the Client Domains of the above architecture via the User Interface Layer.

The basic flow of the app and can be enlisted as follows:

1. The user searches for a doctor and the app displays the distance-based list of the same. The user can then call the doctor to contact and receive treatment. The search functionality sorts out the list based on distance for all healthcare facilities that it provides.

Figure 3 explains the workflow diagram explaining the flow of the app and the subsequent data flow.

1. The user searches for medical shops and medicines. If medicines are available, then the list of shops that have the stock of that medicine is displayed. If not, then the flow moves to search for the doctor to suggest for his advice on the prescribed medicine.
2. The user can directly dial the contact as the phone number is provided and captured during the registration process.

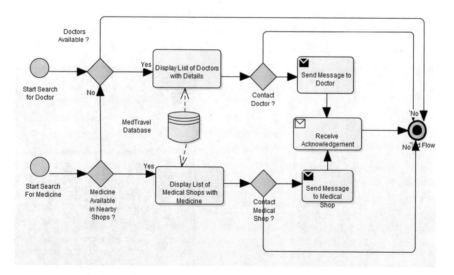

Fig. 3 Workflow diagram

4 Evaluation

The app makes use of Google Maps for displaying the current location of the user as well as locations of all medical shops, hospitals, and doctors. The OS target version used to compile the project is API Level 23 (OS Version 6.0 Marshmallow). The minimum Android OS version required for this project is API Level 16 (OS Version 4.1.1 Jelly Bean). The Android studio version used is 2.1. The app has been developed and the following section evaluates the work along with appropriate screenshots. The next section discusses the app screenshots for evaluation.

In our proposed app to enable integration with maps and show location of doctors and healthcare facilities, we have integrated with Google Maps API to enable location advantages for the users. This has enabled lesser development time and is highly recommended for its reusability

Figure 4 displays the App Home Page with the modules.

On clicking the 'doctor module,' the doctor search results are shown. Figure 5 displays the search results of the doctor search when the user invokes the search in ascending order of distance. For example, the closest doctor Dr Ghosh is displayed first, followed by the next doctor, Doctor Banerji, and so on. The list also displays their timing, address, and specialization

On clicking the Plus icon on the remote right side, the Doctor Registration Page opens up. In this screen, the doctor can register with the app by submitting details about his chamber location, specialization, availability, slot timing, etc. On clicking the medical shops module, the links for shops and medicines open up.

Figure 6 shows the links for searching medicine shops and medicines.

Fig. 4 App home page

Fig. 5 Search results of nearby doctors

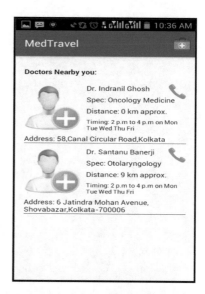

Fig. 6 Search medical shops

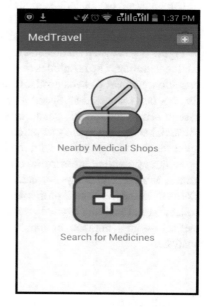

Fig. 7 Search results of
nearby hospitals

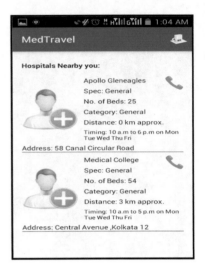

On clicking the hospitals module, the hospital search results are displayed. Figure 7 shows the search results of the nearby hospitals. On clicking the Plus icon on the remote right side, the Hospital Registration Page opens up.

Figure 8 screenshot on hospital registration fields displays the interfaces for hospital registration. The hospital is allowed to register with relevant details like time of operation, number of beds available, address, contact numbers, and specializations (in app, it is displayed as Spec.) available, etc. Time of operation is displayed for ease of access for specific doctor on the basis of particular situation, although this research handles emergency situation mostly. Figure 9 displays the message received upon successful registration with the app.

Besides providing information on the move and enabling patients and doctors to access, analyze, and utilize this data, this also gives opportunities to provide a social platform to the healthcare professionals. Higher end apps even go further for data collection and analytics to yield greater benefits for example monitoring of healthcare metrics—a large numbers of apps monitor biometrics and provide relevant data for analysis.

5 Conclusion

The paper researches on the effect and influence of mobile technology in healthcare. It formulates the basic and urgent healthcare needs into tangible requirements of the end user and patients. These requirements extend into medical amenities including doctors, medical shops, and medicines. The requirements form the basis on which the prototype healthcare mobile app has been developed. The app utilizes Google

Fig. 8 Hospital registration

Fig. 9 Succesful registration notification

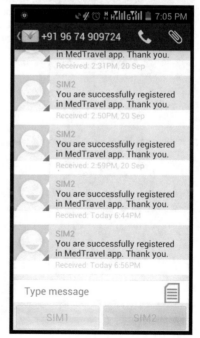

Maps and associated mobile technologies to provide basic functionalities including search and access for medical facilities.

In future, with today's advanced technology, mobile healthcare support to mobile people will be definitely established. Further, the present research shall be extended for accessing the ambulances in the particular required zone or location in remote area, through the MedTravel app, where the user needs sudden medical care on road.

References

1. Chen, G., Yan, B., Shin, M., Kotz, D., Berke, E.: MPCS: Mobile-Phone Based Patient Compliance System for Chronic Illness Care, pp. 1–7,13–16 July 2009
2. Modi, K., Mohanty, R.B.: M Health: Challenges, Benefits and Keys to Successful Implementation. Accessed from https://www.infosys.com/industries/insurance/white-papers/Documents/health-challenges-benefits.pdf on 27-09-2016
3. He, D., Kumar, N., Wang, H., Wang, L., Raymond Choo, K.-K., Vinel, A.: A provably-secure cross-domain handshake scheme with symptoms-matching for mobile healthcare social network. IEEE Trans. Dependable Secure Comput. (Volume: PP, Issue: 99)
4. West, D.M.: Improving Health Care through Mobile Medical Devices and Sensors. Accessed from https://www.brookings.edu/wp-content/uploads/2016/06/West_Mobile-Medical-Devices_v06.pdf on 27-09-2016
5. Das, A., Mitra, P., Basu, D.: Provisioning of medical analysis in cloud. In: Proceedings of 10th INDIA COM 2016, pp. 995–999, held on 16–18th March 2016
6. Assuncao, M.D., Calheiros, R.N., Bianchi, S., Netto, M.A.S., Buyya, R.: Big Data Computing and Clouds: Trends and Future Directions. ScienceDirect, pp. 1–13 (2014)
7. Tsai, W.-T., Sun, X., Balasooriya, J.: Service-oriented cloud computing architecture. In: Seventh International Conference on Information Technology, pp. 1–4 (2010)

A Study on Various Database Models: Relational, Graph, and Hybrid Databases

Shubham Gupta, Sovan Pal and Maumita Chakraborty

Abstract Relational database is a popular database for storing various types of information. But due to the ever-increasing growth of data, it becomes hard to maintain and process the database. So, the graph model is becoming more and more popular since it can store and handle big data more efficiently compared to relational database. But both relational database and graph database have their own advantages and disadvantages. To overcome their limitations, they are combined to make a hybrid model. This paper discusses relational database, graph database, their advantages, their applications and also talks about hybrid model.

Keywords Relational database · Graph database · Big data · Hybrid model

1 Introduction

Relational databases (DBs) have been in use since the 1970s. The data produced then was much less. Ninety percent of the data that was ever created was created in the last 2 years alone. That is how fast the amount of data is growing. Today's data can be characterized as densely connected, semi-structured, and with a high degree of data model volatility. Earlier most of the data was structured. Now, it is a combination of structured as well as unstructured data. Unstructured data includes audio files, video files, pictures, etc.

As we all know, relational DBs are ideal for structured data: the data which follows a schema and order and can be easily placed inside the tables consisting of rows and columns. But of unstructured data, relational database management system (RDBMS) does not perform well enough. Thus, not only SQL (NoSQL) was

S. Gupta (✉) · S. Pal · M. Chakraborty
Institute of Engineering and Management, Kolkata, India
e-mail: gupta.shubham@hotmail.com

S. Pal
e-mail: sovanpal163@gmail.com

M. Chakraborty
e-mail: maumita.chakraborty@iemcal.com

© Springer Nature Singapore Pte Ltd. 2020
M. Chakraborty et al. (eds.), *Proceedings of International Ethical Hacking Conference 2019*, Advances in Intelligent Systems and Computing 1065, https://doi.org/10.1007/978-981-15-0361-0_11

introduced to tackle this problem. It is mainly of four types: key value, document based, column-based, and graph based. Among these, graph-based DBs have shown enormous potential. Big companies have adopted graph DBs like Twitter (FlockDB) and Facebook (TAO Graph Data Store).

Since both graph databases (DBs) and relational databases (DBs) have their own advantages and disadvantages, it is not the goal to absolutely drop out one of the DBs from the plan, but to integrate both of them such that they enhance their strengths and compensate their weaknesses. In other words, we need to create a hybrid model (suggested to be a combination of both relational and graph models) based on the requirements. Some such hybrid models have already been proposed by researchers round the globe.

2 Models

A database model provides the logical structure of a database and also determines how data may be stored or organized. The most common one is the relational model. In this section, we are going to talk about two well-known database models, namely relational and graph models.

2.1 Basic Understanding of Relational Model

The relational model was invented by Edgar F. Codd as a general model of data and subsequently promoted by Chris Date and Hugh Darwen among others. It first came to the force in the 1970s with Codd's relational model of data [1].

A relational database consists of tables that are linked to each other in a meaningful way. A table consists of several rows and columns. Each column is called an attribute, and each row is called a record. Every row has a key to uniquely identify it. Every column and every row in a table are unique. Tables can be connected by making joins [2] and using foreign keys [3]. Relational model has got its name from the close correspondence between table and mathematical concept of relation.

This type of database generally uses Structured Query Language (SQL) as the query language. Transactions with these databases follow the ACID properties [4]. They include:

1. **Atomicity**: Either none or all operations of a transaction are executed.
2. **Consistency**: A database should be consistent before and after a transaction.
3. **Isolation**: No transaction should affect the existence of any other transaction.
4. **Durability**: All latest updates should be present in the database even if system fails.

There is another very important and commonly used concept in database management system, named normalization. It is a process of organizing the data in the

Fig. 1 Fully normalized relation related to students participating in different activities. There are three tables without any redundancy

Students Table

Student	ID*
John Smith	084
Jane Bloggs	100
John Smith	182
Mark Antony	219

Participants Table

ID*	Activity*
084	Tennis
084	Swimming
100	Squash
100	Swimming
182	Tennis
219	Golf
219	Swimming
219	Squash

Activities Table

Activity*	Cost
Golf	$47
Sailing	$50
Squash	$40
Swimming	$15
Tennis	$36

database and is used to minimize redundancy from a relation or set of relations. In this process, the larger table is often divided into smaller tables and they are linked up using relationships [5].

Figure 1 shows a fully normalized relation [6] with no redundancy. As we can see here, ID and activity act as foreign key to connect two different tables.

2.2 Basic Understanding of Graph Model

Graph database is a type of NoSQL database [7]. Graph database consists of nodes and edges like a graph. Node is an entity, and sometimes, it also represents a purpose. Edges are used to represent the relationship. Nodes are connected using relationship. Graph model has a path through which we can traverse from one node to another node. There are two properties of graph database, namely graph storage and graph processing.

Graph Storage: This is required for storing and managing graphs. Specific storage options need to be designed specifically, as the storage required for relational or object-oriented databases is slower.

Graph Processing Engine: As graph databases need to deal with billions of nodes, the capability of fast data exploration and distributed parallel computing is required. It should support both low-latency online query processing and high-throughput offline analytics [8].

Figure 2 shows an example of graph database model consisting of different entities and the relationships between them.

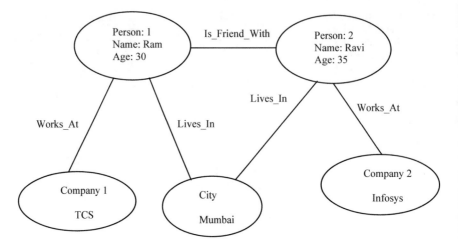

Fig. 2 A graph database model where different vertices of the graph represent different entities and relationships are represented by edges

There are various use cases of graph database, like it is used in fraud detection, master data management, identity, and access management. Companies like Walmart created a recommendation engine with the help of graph database.

Graph database is used in route finding, i.e., going from one point to another.

Tech giants Facebook and Google built a search engine with the help of graph database. In 2012, Google launched 'Knowledge Graph' [9], and in 2013, Facebook introduced 'Graph Search' [10] to generate relevant information.

Knowledge graph puts the information together to create interconnected search results in more efficient manner. Using the knowledge graph, users can get more relevant information [9].

Facebook has many users worldwide, so it has to maintain big data. Now to retrieve proper information from large database and give relevant results to the user are challenging. So Facebook takes the help of graph-based searches, which allows to filter the result using some criteria [10].

For different advantages of graph database, many organizations focus on this area.

In the next section, we will be discussing the relative advantages of graph DB over relational DB and vice versa.

2.3 Advantages of RDBMS Over NoSQL/Graph-Based NoSQL

In this section, we have mentioned some of the important properties of RDBMS which are considered to be advantageous over graph database system [11].

1. Normalized data: No anomalies, less redundancy, every data is stored only once.
2. Relational schema: We always know what columns exist in a row.
3. Fixed data types: Well-defined data types for each column to clearly define which type of data is expected.
4. Authentication and security: Much better in RDBMS than NoSQL.
5. Dominant in the market: Most companies still use RDBMS, and there are more professionals available in this field.
6. ACID: Atomicity, consistency, integrity, and durability properties are maintained which may not be maintained in NoSQL.

2.4 Advantages of Graph DBs Over RDBMS

Here, we have discussed the advantages of graph DBs over that of relational DBs [11].

1. Relationships (edges) between data items (vertices) are given much more importance here.
2. The query is 10 to 100 times shorter than average SQL queries.
3. Data representation through graphs is much more intuitive. Whatever the management plans with a use-case diagram can be easily implemented in exactly the same way in the graph.
4. It is easier to horizontally scale than RDBMS. Horizontal scaling is done by adding more machines into the pool of resources.
5. In contrast to relational databases, where the query performance on data relations decreases as the dataset grows, the performance of graph databases remains relatively constant.
6. Since graph model is used to store the data, maintaining the database is easier compared to relational database and it will also allow us to expand the model in the future.
7. In terms of searching, graph-based search is faster than basic search.
8. Graph-based search returns more relevant information compared to basic search which involves only pattern recognition.

3 Sample Use Cases of Graph Databases

3.1 Graph DB in Networks

Networks are nothing but graphs. To explain this further, we can always visualize any network as a group of nodes/vertices and the interconnections/edges between them.

Thus, graph databases are becoming obvious standards for telecommunications, network, data center, as well as cloud management applications. Graph databases not only store network configurations, they can also generate real-time alerts during possibilities of infrastructure failures. As a result, the time to resolve these problems gets reduced from almost hours to seconds. Hence, we can obviously say that graph databases stand out at storing, querying, and also modeling network data [12].

It has been well-observed that by using a Neo4j graph database, a cohesive view of all computers, network devices, connections, services, applications, and users can be easily created. Neo4j not only maps and manages a network proactively (lists the assets and their deployment in a network, picturize dependencies among various network components, identify bottleneck and risk factors in a network, estimate latency between network nodes), but also increases network visibility (identifying network elements specific to customer needs, identifying applications and services affected because of failures of network devices, identifying paths that may lead to bottlenecks, identifying points where faster connections or devices may be introduced, and also the overall sustainability of the network equipments and topology) as well as controls network growth (graph properties can be easily modified to reflect new network characteristics introduced while upgrading and replacing devices).

3.2 Graph DB in Big Data

Big data generally relied on Structured Query Language (SQL), language of RDBMS, to communicate with a database. This communication between tables slows down when huge and irregular datasets are used. As data keeps on growing, SQL model becomes insufficient in dealing with relationships between different datasets. The graph database model focuses on the relationships of different nodes. Instead of just finding out the value of specific data, the value of relationship between data needs to be understood for any organization. The NoSQL database model can be much more efficient in terms of finding these data connections [13, 14].

4 Previous Work on Hybrid Models of Database

Silvescu et al. [15] introduced the concept of graph databases. In his paper [16], Jaroslav Pokorny has explained the drawbacks of graph database including lack of maturity, functionality restrictions, need for proper benchmarking, big analytics requirements, restrictions during designing, etc. Researches also tried migrating whole relational DBs into graph databases which had many technical hurdles, such as developing algorithms to optimally and accurately transfer data [17, 18]. So, researchers focused on optimally including both databases into hybrid databases which have the best of both worlds (graph and relational).

Jeff Shute has worked on a hybrid database called F1 which has the scalability and availability of a NoSQL database, and also consistency and usability of relational databases [19].

Another hybrid database system approach is designed by Blessing E. James and P. O. Asagba for the storage and management of big data. Big data contains a lot of unstructured data which cannot be put inside a pre-defined schema. Hence, they built a hybrid model consisting of MongoDB and MySQL to help in storing the big data [20].

J. O. Little developed Grapht which provides an intermediate query processing layer between RDBMS and in-memory graph store. When user hits query, the query processor divides it into row centric sub-queries for relational DBs and graph-centric sub-queries for the graph handler [21].

Since SQL is a standard query language, it has several benefits. To reap the benefits, Luis Ferreira tried to build layer between SQL code and the interpreter and the actual NoSQL database underneath them. This model used SQL queries to query on a NoSQL system [22].

5   A Sample Hybrid Model

The relational databases cannot be completely migrated to a NoSQL platform or a graph-based platform because it will be very tedious and time consuming to do so. Most of the data that is stored today is in relational format. Moreover, the relational model has too many advantages to completely remove it from proposed models. So, it is better to come up with a hybrid system [23, 24] involving both relational and NoSQL models which can cancel out each other's flaws and enhance each other's strength in the required areas.

As SQL (query language used in RDBMS) is much more common than Cypher (a declarative graph query language used in graph DB) or another graph-based querying language, it would be easier and more practical if a model is made so that users can query in SQL rather than Cypher. Also, SQL will be required to query the relational part of the database and Cypher code will be required to query the graph-based database. So, the SQL queries may be converted into Cypher by using certain algorithms for accessing graph-based databases. The SQL queries are generally large and take more space to store and read. So, Cypher codes will reduce the size of the queries too.

To store the data, we know graph-based models are better for unstructured data and relational models are better for structured data. So, a classifier algorithm may be used to classify the structured and unstructured data and store them in relational database and graph database, respectively.

A sample hybrid model which combines the concepts of both relational DB and graph DB has been shown in Fig. 3. It has the scope of selecting any one model based on the type of data.

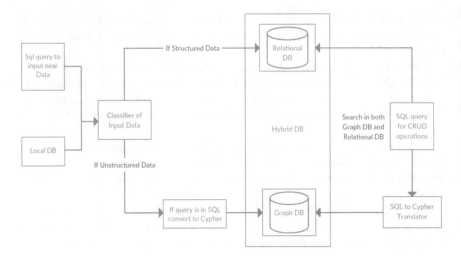

Fig. 3 A sample hybrid model showing the scope of selecting both relational DB and graph DB based on the type of data to be stored

6 Conclusion

In this paper, we have discussed relational as well as graph-based database models. Relational databases have a lot of strengths. There is a good reason that most databases even in the modern times are relational and not NoSQL databases. Still, graph databases are evolving. The relative advantages of one over the other have been discussed along with their applications. We can conclude by stating that relational models will always be relevant in modern DBMS. This paper also talks about hybrid DBs, which are considered to be the future where certain elements of NoSQL can improve the usability and storability of the databases. Many such hybrid database models do exist and many others will come to existence with more and more research and development in this field.

References

1. Codd, E.F.: A relational model of data for large shared data banks. Commun. ACM **13**(6), 377–387 (1970). https://doi.org/10.1145/362384.362685
2. SQL Join: https://www.dofactory.com/sql/join. Last accessed 15 July 2019
3. SQL Foreign Key, https://www.tutorialspoint.com/sql/sql-foreign-key. Last accessed 15 July 2019
4. DBMS Transaction: https://www.tutorialspoint.com/dbms/dbms_transaction. Last accessed 15 July 2019
5. Normalization: https://www.javatpoint.com/dbms-normalization. Last accessed 24 July 2019
6. Database normalization, https://en.wikipedia.org/wiki/Database_normalization. Last accessed 15 July 2019
7. Graph Database: https://en.wikipedia.org/wiki/Graph_database. Last accessed 15 July 2019

8. Graph Engine: https://www.graphengine.io/. Last accessed 15 July 2019
9. Knowledge Graph: https://en.wikipedia.org/wiki/Knowledge_Graph. Last accessed 15 July 2019
10. Introducing Graph Search Beta: https://newsroom.fb.com/news/2013/01/introducing-graph-search-beta/. Last accessed 15 July 2019
11. Relational Databases vs. Graph Databases: A Comparison, https://neo4j.com/developer/graph-db-vs-rdbms/. Last accessed 15 July 2019
12. Managing Network Operations with Graphs: https://neo4j.com/business-edge/managing-network-operations-with-graphs/. Last accessed 15 July 2019
13. Difference between SQL and NoSQL—GeeksforGeeks: https://www.geeksforgeeks.org/difference-between-sql-and-nosql/. Last accessed 15 July 2019
14. Graph Databases in Big Data Analytics: https://www.cleverism.com/graph-databases-effective-big-data-analytics/. Last accessed 15 July 2019
15. Silvescu, A., Caragea, D., Atramentov, A.: Graph database. artificial intelligence research laboratory, Department of Computer Science, Iowa State University [Online]. http://people.cs.ksu.edu/~dcaragea/papers/report.pdf (2012)
16. Pokorny, J.: Graph databases: their power and limitations. In: IFIP International Conference on Computer Information Systems and Industrial Management, pp. 58–69. Springer, Cham (2015)
17. Virgilio, R.D., Maccioni, A., Torlone, R.: Converting relational to graph databases. In: Proceedings of ACM First International Workshop on Graph Data Management Experience and Systems, vol. 1, pp. 1–6 (2013). http://doi.acm.org/10.1145/2484425.2484426
18. Bordoloi, S., Kalita, B.: Designing graph database models from existing relational databases. Int. J. Comput. Appl. **74**(1), 25–31 (2013)
19. Shute, J., et al.: F1: a distributed SQL database that scales. Proc. VLDB Endowment **6**(11), 1068–1079 (2013)
20. James, B.E., Asagba, P.O.: Hybrid database system for bigdata storage and management. Int. J. Comput. Sci. Eng. Appl. (IJCSEA) **7**(3/4) (2017)
21. Little, C.J.O.: Grapht: a hybrid database system for flexible retrieval of graph-structured data. In: Master's thesis, University of Cambridge, Emmanuel College, Cambridge, United Kingdom (2016)
22. Ferreira, L.: Bridging the gap between SQL and NoSQL. In: A state of art report, Universidade do Minho, pp 187–197 (2011)
23. Hybrid Databases: Combining relational and NoSQL. https://www.stratoscale.com/blog/dbaas/hybrid-databases-combining-relational-nosql/. Last accessed 15 July 2019
24. Vyawahare, H., Karde, P., Thakare, V.: A hybrid database approach using graph and relational database. In: International Conference on Research in Intelligent and Computing in Engineering (RICE) (2018)

Microcontroller-Based Automotive Control System Employing Real-Time Health Monitoring of Drivers to Avoid Road Accidents

Mohuya Chakraborty and Arup Kumar Chattopadhyay

Abstract This paper is aimed at preventing car accidents by managing four signifi-cant aspects related to car drivers. Firstly, it has been observed that 10–30% of road accidents are related to drowsiness of the drivers mainly at night or at drunken state. Detecting the drowsiness in drivers and alerting him can improve the safety on roads. The system could also measure alcohol molecules in driver's breath and automati-cally halt the car if the legal drinking limit is exceeded. Secondly, accidents occur due to medical emergency conditions of the drivers. A system that continuously monitors the health of drivers can effectively reduce accidents. Thirdly, the designed system would continuously monitor the distance of the vehicle from obstacle by the use of Light Detection and Ranging (LIDAR). The LIDAR upon detection of the obstacle would warn the driver as well as decrease the speed of the vehicle and will stop the vehicle when reaches a certain distance of the obstacle by actuating the braking system and ignition system. Fourthly, the system would also monitor lane changing to assist drivers to ensure that their vehicles are within lane constraints when driving, so as to make sure traffic is smooth and minimize chances of collisions with other cars in nearby lanes. We propose to implement the system using microcontroller and a few numbers of desired heartbeat, ultrasonic and breathe-based sensors to detect irregular heartbeat in case of medical emergency of the driver, alcohol content of the driver and distance of the vehicle from the obstacle, respectively. For brain activity, we have designed an artificial neural network model on field-programmable gate array (FPGA) to detect drowsiness of the driver. The initial phase of the research work has been conducted in the laboratory environment where all the electronics cir-cuit with sensors and microcontroller have been built up and tested at the laboratory environment on human subjects. In the second phase, the electronic circuits were integrated into the car. The testing of the project has been performed in a controlled environment at workshop. Results show the efficiency and benefit of the proposed research work.

M. Chakraborty (✉) · A. K. Chattopadhyay
Institute of Engineering and Management, Kolkata 700091, India
e-mail: mohuyacb@iemcal.com

A. K. Chattopadhyay
e-mail: arup.chattopadhyay@iemcal.com

© Springer Nature Singapore Pte Ltd. 2020
M. Chakraborty et al. (eds.), *Proceedings of International Ethical Hacking Conference 2019*, Advances in Intelligent Systems and Computing 1065, https://doi.org/10.1007/978-981-15-0361-0_12

Keywords Microcontroller · Automotive · Control · Artificial neural network · Heart rate · Drowsiness · Vehicle safety · Alcohol detection · Smart vehicle

1 Introduction

One of the major causes of road accidents is deterioration in physical and mental condition of the driver. A safety system based on biosensors can be very vital to prevent road accidents. Current research works used wearable sensors and cameras to detect different physical health conditions. But the mental health condition is mostly ignored. Different research works use different kinds of sensors for real-time health monitoring of driver. This real-time data are analyzed such that the action can be taken to trigger some emergency system to alert the driver. For detecting the alcohol consumption, the traditional approach is to use MQ3 sensor to detect alcohol concentration on breath. But the sensor may also catch the input from co-passengers. So, a specific measure has to be taken by the system. Further, the effect of alcohol on the driver can be identified by driver's head movement. In the previous works, drowsiness of drivers has been normally identified by the eye blink and head movement. The eye blink and head movements are detected by the camera installed in front of the driver. But, it requires image processing which is time consuming and ineffective in real-time situation. We rather propose to use simple motion sensors to capture head movement data above threshold level and analyze brain waves from a brain–computer interface (BCI) system mounted on the head of the driver, to identify the eye blink rate.

The use of BCI system proposed here is most important to monitor the drowsiness in the driver. Intelligent software, a subsystem of BCI will be trained to classify different mental state of the driver. A trained system can find out the drowsiness in the driver very effectively in real time. We have proposed four different prevention mechanisms to build the smart car. Firstly, in case of alcohol consumption by the driver, the car will be locked until and unless the former is replaced by some sober driver. Secondly, in case of abnormal heart rate or heart rate not detected for a stipulated time, the car will be slowly stopped and SMS will be sent to the registered mobile numbers giving location by GPS for help. Thirdly, in case of drowsiness, the driver will be alerted by activating the emergency system. Fourthly, by using either ultrasonic or LIDAR sensor, detection of any obstacle in the proximity of the vehicle would warn the driver as well as decrease the speed of the vehicle.

In today's world, the major cause of road accidents can be attributed to alcohol consumption and physical and mental health condition of the driver. The main aim is to integrate the automobile parts of the car with a bio-system-based structure to alter the driver's mental and physical state in those situations and start accident prevention mechanisms. Our proposed project is aimed at preventing car accidents by managing three significant aspects related to car drivers. Firstly, it has been observed that 10–30% of road accidents are related to drowsiness of the drivers mainly at night or at drunken state. Detecting the drowsiness in drivers and alerting them can improve the safety on roads. The system could also measure alcohol molecules in driver's breath and automatically halt the car if the legal drinking limit is exceeded. Secondly, accidents occur due to medical emergency conditions of the drivers. A system that continuously monitors health of drivers can effectively reduce accidents. Thirdly, the designed system would continuously monitor the distance of the vehicle from obstacle by the use of ultrasonic sensor.

The ultrasonic sensor upon detection of the obstacle would warn the driver as well as decrease the speed of the vehicle and will stop the vehicle when reaches a certain distance of the obstacle by actuating the braking system and ignition system. We propose to implement the project using microcontroller and a few numbers of desired heartbeat, ultrasonic and breath-based sensors to detect irregular heartbeat in case of medical emergency of the driver, alcohol content of the driver and distance of the vehicle from the obstacle, respectively. For brain activity, we would design an artificial neural network model on field-programmable gate array (FPGA) to detect drowsiness of the driver. A lightweight headgear consisting of a number of brain wave sensors can be mounted on driver's head. The brain signal transmitted to FPGA would analyze different types of brain waves (such as Alpha, Beta, Theta, Gamma and Delta) to identify the drowsiness/sleepiness of the car driver. The sensors on headgear can also detect variation in eye blink to help to detect the same. The state of autonomous nervous system also has its reflection on heart rate variability (HRV). A number of heart rate sensors mounted at specific positions at driver's body can also help to detect fatigue and drowsy conditions in the driver. Whenever drowsiness is detected by the system, the microcontroller activates a buzzer to alert the driver. If the buzzer is not interrupted by the driver within 30 s, it activates the hazard lights on the car and starts slowing down the car to entire halt. Sensors mounted on driver's body would continuously monitor heart rate and ECG signals to identify heart attack and other health-related problems of driver during the state of driving. Upon detection and analysis of these signals, information in the form of SMS would be sent to predefined five mobile numbers of relatives or friends by using GSM module and bring the car to stop entirely.

If the drowsiness is caused by alcohol consumption, it can be detected by alcohol gas sensor, and the car will not start until the drunken driver is changed with a sober driver or his alcohol effect reduces to RTO allowed level. The system will incorporate different sensors with the vehicle and will use different electromechanical actuator to get the desired results as per the sensors and processor requirements.

The organization of the paper is as follows. After the introduction in Sect. 1, the background study of the research work is given in Sect. 2. Section 3 presents the overview of the proposed system. Section 4 concludes the paper with glimpses of future work.

2 Background Study to Conduct the Research Work

2.1 International Status

Biofeedback headsets measure heart rate, blood pressure and body temperature, and the most important parameters of human body may be detected by using Arduino microcontrollers and sensors [1]. In [2], the designers used heart rate sensor module to detect heart rate. The sensor module contains an IR pair which actually detects heartbeat from blood.

In a similar way, we can record and measure brain activities by using tiny, low-cost and easy-to-use wearable equipment [3]. Measurement of brain waves for different types of activities occurring in the human brain may be accomplished by using EEG, which can have too much complexity depending upon the region of the brain where the sensors are placed [4]. The different brain waves that are generated at different frequencies and for different activities are as follows:

- Delta (0.1–3.9 Hz)—during deep and dreamless sleep
- Theta (4–7.9 Hz)—during intense dream sleep (REM)
- Alpha (8–13.0 Hz)—during tranquil state, listening to music
- Beta (14–30 Hz)—during awake state, talking
- Gamma (31 Hz+)—very much attentive.

If the alcohol content of the drivers exceeds the legal limit, cars may be caused to stop by using blood alcohol sensor technology [4]. In [5], researchers used human sweat to design a biochemical sensor platform suitable for biosensing of ethyl glucuronide (EtG) present in the sweat.

Vehicle automation features are becoming more and more important in the field of sophisticated driver support systems in order to increase the safety and comfort of the vehicle [6–8]. Different sensors and network structure are being used to make the vehicle autonomous and smart; numbers of different works are reported in the literature using broad sensing methodology including tracking of gaze, blink behavior, facial tracking, eye closure, EEG measures, lane tracking, vehicle lateral position, steering wheel input, pupillometry measures, mayo pupillometry system, ECG, EOG skin, pulse and oxygen saturation in blood and steering grip change. Similarly, traffic accidents throughout the world mainly occur due to drunk driving.

A drive by wire mechanism was also studied to understand the safety of the vehicle by Bergmiller [9]. A behavioral-based control system of an autonomous vehicle was developed by Pack et al. [10]. While operating autonomously, the vehicle will be alert of its environment, particularly the site and character of terrain features that may be an obstacle. Because these terrain features can represent hazards to the vehicle, the vehicle preferably compensates by altering its trajectory or movement. Driver inattention is another factor in most traffic accidents.

A highly efficient system was developed by Seshadri et al. [11] for early detection and warning of dangerous automobile driving particularly associated with drunk driving by the use of a mobile phone, accelerometer and orientation sensor. Driver sleepiness is another problem and causes most of the road accidents. Researchers have tried to detect driver sleepiness by using vehicle-based measures, physiological measures and behavioral measures [12]. A blink detection-based camera system was studied and showed a good percentage of detection of drowsiness in real road trials. However, problems persist with poor light conditions and for spectacled persons [13, 14].

Over the past few decades, researches using several methodologies have been undertaken to accurately determine the state of mind of the driver. Some of these measures are driver biological measures, subjective report measures, driver physical measures, driving performance measures and hybrid measures. The hybrid measures give more consistent solutions compared with single driver physical measures or driving performance measures, because the hybrid measures reduce the number of false alarms and maintain a high detection rate, which encourage the acceptance of the system [15].

2.2 National Status

Sleepiness of drivers is the major reason of road accidents causing casualties and financial loss in recent years. Researchers have developed image processing-based drowsiness detection system. Image processing techniques are used to process the images of the driver's facial and head movement for identifying driver's current state, viz. drowsiness. Sensors are used for detection of driver's alcohol consumption. The number of road accidents might then be avoided if an alert is sent to a driver that is deemed drowsy. Camera-based drowsiness measure has a major contribution in this field [16].

In one project, researchers have also developed a means that can detect driver tiredness and set off an alarm in the form of a flashing light or siren on the dashboard or even an order to shut down the engine and bring the car to a halt. Driver performance is detected based on many factors like brain and muscle activities as well as body pressure distribution [17]. In yet another project, researchers have developed a system to keep the vehicle secure and protect it by the occupation of the intruders.

This project involves measurement of the eye blink using IR sensor and controlling accident due to unconsciousness through eye blink. Here, one eye blink sensor is fixed in vehicle where if anybody loses consciousness, it is indicated through alarm [18]. Kodi and Manimozhi [19] developed a customized lane detection algorithm to detect the curvature of the lane for autonomous vehicle.

A ground truth labeling toolbox for deep learning is used to detect the curved path. A non-intrusive system which can detect fatigue of the driver and give a timely warning was developed by Sing et al. [20]. Driver fatigue detection based on tracking the mouth and monitoring and recognizing yawning was proposed by Saradadevi and Bajaj [21]. A system for onboard monitoring the loss of attention of an automotive driver, based on PERcentage of eye CLOSure (PERCLOS) was developed by Dasgupta et al. [22].

3 Proposed System

The proposed system consists of sensors that collect the driver's physical conditions and different information about the car itself. Then, the collected information will be processed on the processing units like Arduino and FPGA module. After processing and decision making, the actuation will be performed on the car's braking system and other parts to control the car. The block diagram and flowchart of the system are shown in Figs. 1 and 2, respectively.

Once the driver tries to start the ignition process of the car at first, the alcohol detection module would be activated. The positioning of the sensor modules is shown in Fig. 1. If it detects alcohol consumption beyond the legal limit, then the ignition process will stop immediately. Otherwise, the car would get started. The alcohol detection module will be still active. The tilt sensors module can detect the tilting of the head which is also considered along with alcohol consumption. The heart

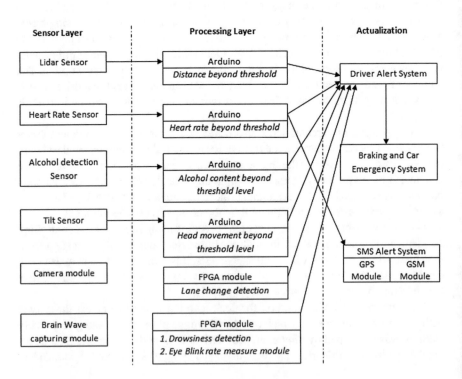

Fig. 1 Block diagram of the proposed system

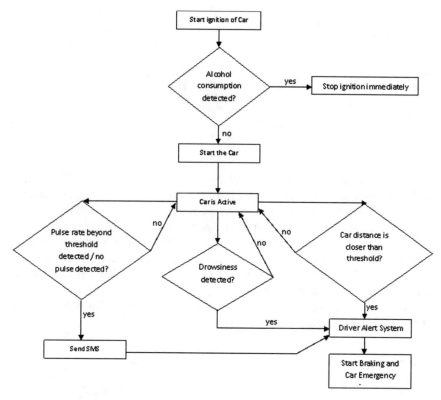

Fig. 2 Flowchart of the system

rate/pulse rate sensors are mounted onto the steering which will measure the BPM of the driver (as shown in Fig. 3). If the BPM is more than the threshold or no pulse is detected for a stipulated time, it activates the alert system and slowly stops the car. It also sends SMS to three registered mobile numbers informing the emergency situation of the driver.

When the car is running, the brain gear mounted over the head of driver (as shown in Fig. 4) sends the brain wave signal to the intelligence system. In our scheme, we

Fig. 3 Positioning of alcohol detection and pulse/heart rate measuring sensor

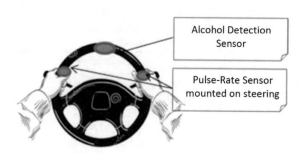

Fig. 4 Positioning of brain
gear to capture brain signals

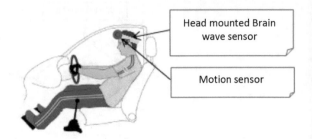

proposed implementation of our intelligent system with FPGA module. The system
would analyze and detect rate of eye blink and drowsiness pattern in the driver. If
drowsiness is detected, the driver alert system would be first activated. If the driver
does not respond in the stipulated time, the system will bring the car to a halt.

The LIDAR sensors at different positions (the positioning is shown in Fig. 5)
continuously monitor car's distance from the neighboring cars and other obstacles. It
activates the alert system if the distance is below the threshold. The camera mounted
on the car helps to detect the lane change.

With reference to Fig. 1, various types of sensors and devices that have been used
for experimental purpose are given in Table 1.

Fig. 5 Positioning of brain
gear to capture brain signals

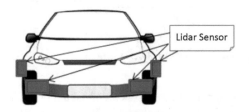

Table 1 Sensors and devices
used in the experimental
environment

Sensor	Utilization
Pulse sensor (heart rate detector)	To measure BPM (pulse rate)
NeuroSky brain wave kit	To measure drowsiness To measure eye block rate
Tilt sensor	Head movement
Alcohol detector sensor (MQ3)	To measure alcohol particles in breath
LIDAR sensor	To measure distance of any obstacle
PI camera	To identify lean change

4 Conclusion

This paper has discussed about a novel idea of integrating an automobile with microcontroller-based bio-based system structure for real-time health monitoring of drivers to avoid road accidents to ultimately produce a smart car. The proposed system has been implemented in controlled environment in the workshop and has been built up and tested at the laboratory environment on human subjects. The results proved to be successful in terms of four different prevention mechanisms built on the smart car like (1) alcohol consumption detection of the driver wherein the car would not move, (2) abnormal heart rate detection of the driver in which case the car would come to a halt slowly and SMS would be sent to registered mobile numbers giving location by GPS for help, (3) drowsiness detection of the driver in which case the emergency system would be activated alerting the driver and (4) obstacle detection in the proximity of the car and lane change detection to warn the driver for speed decrease. In future, we would like to test the smart car on the road by taking into account actual road traffic, obstacle and other related factors for real-time implementation.

References

1. Mallick, B., Patro, A.K.: Heart rate monitoring system using finger tip through arduino and processing software. Int. J. Sci. Eng. Technol. Res. (IJSETR) 5(1), 84–89 (2016)
2. https://circuitdigest.com
3. Kannan, V.R., Joseph, K.O.: Brain controlled mobile robot using brain wave sensor. In: International Conference on Emerging Trends in Engineering and Tehnology Research. Published in IOSR Journal of VLSI and Signal Processing No., pp. 77–82, e-ISSN: 2319 –4200, p-ISSN No.: 2319 –4197
4. https://www.diygenius.com
5. Kim, J., Jeerapan, I., Imani, S., Cho, T.N., Bandodkar, A., Cinti, S., Mercier, P.P., Wang, J.: Nonnvasive alcohol monitoring using a wearable tatto-based ionotophoretic- biosensing system. ACS Sens. 1(98), 1011–1019 (2016)
6. https://www.dadss.org
7. Selvam, A.P., Muthukumar, S., Kamakoti, V., Prasad, S.: A wearable biochemi cal sensor for monitoring alcohol consumption lifestyle through Ethyl glucoronide (EtG) detection in human sweat. Sci. Rep. 6:23111 (2016). Published online 21 Mar 2016. https://doi.org/10.1038/srep23111
8. Lotz, F: System architectures for automated vehicle guidance concepts. In: Automotive Systems Engineering, pp. 39–61. Springer, Berlin (2013)
9. Bergmiller, P.: Design and safety analysis of a drive-by-wire vehicle. In: Automotive Systems Engineering, pp. 147–202. Springer, Berlin (2013)
10. Pack, R.T., Allard, J., Barrett, D.S., Filippov, M., Svendsen, S.: U.S.Patent No. 9,513,634. U.S. Patent and Trademark Office, Washington, DC (2016)
11. Seshadri, K., Juefei-Xu, F., Pal, D.K., Savvides, M., Thor, C.P.: Driver cell phone usage detection on strategic highway research program (SHRP2) face view videos. In: Proceedings of the IEEE Conference on Computer Vision and Pattern Recognition Workshops, Boston, MA, USA, pp. 35–43 (2015)
12. Sahayadhas, A., Sundaraj, K., Murugappan, M.: Detecting driver drowsiness based on sensors: a review. Sensors 12(12), 16937–16953 (2012)

13. Friedrichs, F., Yang, B.: Camera-based drowsiness reference for driver state classification under real driving conditions. In: Intelligent Vehicles Symposium (IV), San Diego, CA, USA, pp. 101–106. IEEE (2010)
14. Williamson, A., Chamberlain, T.: Review of on-road driver fatigue monitoring devices. NSW Injury Risk Management Research Centre, University of New South Wales (2005)
15. Dong, Y., Hu, Z., Uchimura, K., Murayama, N.: Driver inattention monitoring system for intelligent vehicles: A review. IEEE Trans. Intell. Transpo. Syst. **12**(2), 596–614 (2011)
16. Zadane, M, Jadhav, P., Totre, P., Jagdale, T., Mankhai, S.: Driver drowsiness detection and alcohol detection using image processing. Int. Res. J. Eng. Technol. (IRJET) **4**(5) (2017)
17. https://timesofindia.indiatimes.com/city/chennai/IIT-Ms-wake-up-call-for-tired-drivers/ articleshow/28869401.cms
18. Kusuma Kumari, B.M.: Detect and prevent accident due to driver drowsiness. Ind. J. Comput. Sci. Eng. (IJCSE) **8**(5), 578–583 (2017)
19. Kodi, B., Manimozhi, M.: Curve path detection in autonomous vehicle using deep learning. Preprints 2018, 2018050326. https://doi.org/10.20944/preprints201805.0326.v1
20. Singh, H., Bhatia, J.S., Kaur, J.: Eye tracking based driver fatigue monitoring and warning system. In: Proceedings of India International Conference on Power Electronics (IICPE), New Delhi, India, pp. 1–6. IEEE (2010)
21. Saradadevi, M., Bajaj, P.: Driver fatigue detection using mouth and yawning analysis. Int. J. Comput. Sci. Network Secur. **8**(6), 183–188 (2008)
22. Dasgupta, A., George, A., Happy, S.L., Routray, A., Shanker, T.: An on-board vision based system for drowsiness detection in automotive drivers. Int. J. Adv. Eng. Sci. Appl. Math. **5**(2–3), 94–103 (2013)

Surface Potential Profile of Nano Scaled Work Function Engineered Gate Recessed IR Silicon on Insulator MOSFET

Tiya Dey Malakar, Moutushi Singh and Subir Kumar Sarkar

Abstract In this present analysis, we represent the surface potential profile of horizontally graded binary metal alloy gate (work function engineered gate) recessed source/drain (Re S/D) SOI/SON MOSFET with additional insulator region (I-SOI). The proposed structure is akin to that of the recessed S/D SOI MOSFET with the exception that there is an insulator region of high-k dielectric in between the channel and drain region. The analytical surface potential model has been developed by solving two-dimensional Poisson's equation in the channel region considering appropriate boundary condition with a parabolic potential profile.

Keywords High-k dielectric · Short channel effects (SCEs) · Recessed source/drain (Re S/D) · Work function engineered gate (WFEG)

1 Introduction

With the continuous downsizing of the device configuration into the nanometer region introduces numerous short channel effects (SCEs) such as drain-induced barrier lowering (DIBL), hot-carrier effect (HCE), and high gate current, etc. [1, 2]. To abolish the shortcomings that arise from downsizing the MOS technology, researchers have found fully depleted silicon on insulator (FD SOI) MOSFET as a potential candidate for their superior electrical characteristics such as lower parasitic capacitances, better immunity against radiation, and higher speed of operation [3–5]. However, FD SOI

T. D. Malakar
Department of Electronics and Communication Engineering, RCC Institute of Information Technology, Kolkata, India
e-mail: tiya_biet@yahoo.co.in

M. Singh (✉)
Department of Information Technology, Institute of Engineering and Management, Kolkata, India
e-mail: moutushisingh01@gmail.com

S. K. Sarkar
Department of ETCE, Jadavpur University, Kolkata, India
e-mail: su_sircar@yahoo.co.in

© Springer Nature Singapore Pte Ltd. 2020
M. Chakraborty et al. (eds.), *Proceedings of International Ethical Hacking Conference 2019*, Advances in Intelligent Systems and Computing 1065, https://doi.org/10.1007/978-981-15-0361-0_13

is not fully immune to various short channel effects so the researcher has invented another unique device structure by a little modification of this SOI technology, named as silicon on nothing (SON) [6–8]. Still this SOI/SON structure suffers from threshold voltage instability and DIBL at higher drain bias which can be overcome with work function engineered gate (WFEG) SOI MOSFETs as reported by Deb et al. [9, 10]. To increase the channel conductivity of the conventional MOSFET structure, the recessed SOI structure has been investigated and fabricated by Zhang et al. and Long et al. [11].

2 Analytical Modeling

Figure 1 shows the schematic diagram of layered WFEG Re S/D IR SOI/SON MOSFET structure. The thicknesses of front gate oxide, buried layer, silicon substrate, and channel silicon film are represented by t_f, $t_{box/air}$, t_{sub}, and t_{si}, respectively. Here, L represents the device channel length; source/drain recessed thickness and length of the source/drain overlap region over buried layer are described by the t_{rsd} and $d_{box/air}$, respectively.

The overall work function of the Pt–Ta the binary alloy system associated system with horizontally varying mole fraction can be expressed as follows:

$$\phi_{meff}(x) = (x/L)\phi_b + (1 - x/L)\phi_a \tag{1}$$

Here, ϕ_a and ϕ_b represent the work functions of Pt and Ta, respectively, so that $\phi_{meff}(x) = \phi_b$ at the drain end ($x = L$) and $\phi_{meff}(x) = \phi_a$ at source end ($x = 0$) [10].

Surface Potential Distribution
To obtain the surface potential distribution of our proposed model, two-dimensional

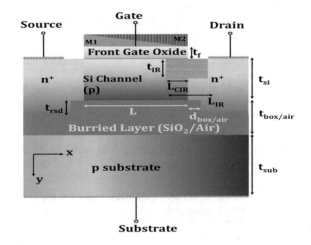

Fig. 1 Cross-sectional view of layered WFEG Re S/D IR SOI/SON MOSFET

Poisson's equation has been solved assuming uniform charge distribution in the thin silicon film region [12].

$$\frac{\partial^2 \phi_i(x, y)}{\partial x^2} + \frac{\partial^2 \phi_i(x, y)}{\partial y^2} = \frac{q N_a}{\varepsilon_{Si}} \quad i = 1, 2 \tag{2}$$

Equation (2) can be resolved by using parabolic potential approximation [13], and the 2D potential profile in the channel region can be expressed as follows:

$$\phi_1(x, y) = \phi_{s1}(x) + m_{11}(x)y + m_{12}(x)y^2 \quad \text{For } (0 \le x \le L - L_{CIR}, 0 \le y \le t_{Si}) \tag{3}$$

$$\phi_2(x, y) = \phi_{S2}(x) + m_{21}(x)y + m_{22}(x)y^2 \quad \text{For } (0 \le x \le L_{CIR}, 0 \le y \le t_{Si}) \tag{4}$$

Here, front interface surface potential in the region I and region II represented by $\phi_{s1}(x)$ & $\phi_{s2}(x)$, respectively, and $m_{11}(x), m_{12}(x), m_{21}(x)$, and $m_{22}(x)$ are the arbitrary coefficient which are function of x only.

(i) *At the front gate, oxide–Silicon channel interface electric field is continuous*:
  Therefore, we have,

$$\left. \frac{d\phi_1(x, y)}{dy} \right|_{y=0} = \frac{\varepsilon_{ox}}{\varepsilon_{-si}} \frac{\phi_{s1}(x) - (V_{gs} - V_{ff}(x))}{t_f} \quad \text{for region I} \tag{5}$$

$$\left. \frac{d\phi_2(x, y)}{dy} \right|_{y=0} = \frac{\varepsilon_{ox}}{\varepsilon_{-effective}} \frac{\phi_{s2}(x) - (V_{gs} - V_{ff}(x))}{t_f} \quad \text{for region II} \tag{6}$$

where front channel interface flat band voltage is represented by $V_{ff}(x) = \phi_{meff} - \phi_S$, and relative permittivity of silicon and silicon dioxide is denoted by ε_{ox} and ε_{-Si}, respectively. $\varepsilon_{-effective}$ is the effective dielectric constant of the region II.

(ii) *At the silicon channel–buried layer interface in region I and region II, electric field is continuous*:

$$\left. \frac{d\phi_1(x, y)}{dy} \right|_{y=t_{si}} = \frac{C_{box1}(V_{Sub\_eff} - \phi_{b1}(x))}{\varepsilon_{-si}} + \frac{C_{rsd1}(V_{S\_eff} - \phi_{b1}(x))}{\varepsilon_{si}}$$

$$+ \frac{C_{rsd2}(V_{D\_eff} - \phi_{b1}(x))}{\varepsilon_{si}} \tag{7}$$

$$\left. \frac{d\phi_2(x, y)}{dy} \right|_{y=t_{si}} = \frac{C_{box2}(V_{Sub\_eff} - \phi_{b2}(x))}{\varepsilon_{-effective}} + \frac{C_{rsd3}(V_{S\_eff} - \phi_{b2}(x))}{\varepsilon_{-effective}}$$

$$+ \frac{C_{rsd4}(V_{D\_eff} - \phi_{b2}(x))}{\varepsilon_{-effective}} \tag{8}$$

Here, effective substrate, drain and source biases are represented by $V_{Sub\_eff} = V_{sub-} - V_t \ln(N_{sub}/n_i)$, $V_{D\_eff} = \underline{V_D} - V_t \ln(N_A N_D/n_i^2)$, and $V_{S\_eff} = V_S$

$- V_t \ln(N_A N_D / n_i^2)$, respectively. $C_{box1} = C_{box2}$ is the buried layer capacitance and other four capacitances arise due to the recessed source/drain regions as proposed by [14] and calculated according to the proposed device structure.

(iii) *Source side potential is given by*

$$\phi_1(0, 0) = \phi_{s1}(0) = V_{blt} \tag{9}$$

(iv) *Drain side potential is given by*

$$\phi_2(L, y) = \phi_{s2}(L) = V_{blt} + V_{ds} \tag{10}$$

(v) *Surface potential at the interface of the two dissimilar channel materials is continuous*:

$$\left. \frac{d\phi_1(x, y)}{dx} \right|_{x=L-L_{CIR}} = \left. \frac{d\phi_2(x, y)}{dy} \right|_{x=L-L_{CIR}} \tag{11}$$

The junction built-in potential is defined by V_{blt}. Using parabolic potential approximation in Eq. (3), the 2D Poisson's equation can be solved and then the first two boundary conditions have been used to calculate the values of coefficient.

Now, substituting the values of these coefficients in Eqs. (3) and (4) and then in Eq. (2), we get the differential equation for front and back surface potentials.

3   Simulation and Result

Analytical modeling of the work function engineered gate (WFEG) recessed S/D IR SOI and SON MOSFETs structures has been proposed. For both the structures, a Ta–Pt binary alloy system is considered as gate electrode with linearly varying work function from 100% Pt (from source side) to 100% Ta (at drain side). The parameters used for the simulations are given in Table 1.

Figure 2 depicts the front surface potential distribution of the proposed structure along the channel lengths. This figure shows that there is a change of surface potential behavior in the IR region, as inserted HfO_2 in the channel region creates depletion region under the gate causing lower trap density there by reducing HCE, and hence, surface potential is modified in this region. Also, the presence of insulator region with a dielectric constant of 22 which is 4–6 times greater than the dielectric constant of SiO_2 causes the potential minima to shift upward resulting reduction in the device threshold voltage. Again, due to the lower value of source to drain potential barrier for electrons in SON structure as compared to SOI structure implies marginally higher potential minima and thereby lower threshold voltage.

In Fig. 3, front surface potential distribution along the channel for various values of drain to source voltages V_{ds} has been depicted. It shows that the surface potential minima independent on the variation of drain bias, demonstrating its better immunity

Table 1 Value of parameters used for simulation

Parameters	Values
N_A	10^{21} m^{-3}
N_{SUB}	10^{21} m^{-3}
$N_{S/D}$	10^{26} m^{-3}
t_{Si}	20 nm
t_f	1.5
$t_{box/air}$	100 nm
L_{IR}	20 nm
L_{CIR}	4 nm
$d_{box/air}$	3 nm
t_{sub}	200 nm

Fig. 2 Front surface potential variation along the channel length of WFEG RE S/D IR SOI/SON structures for $V_{gs} = V_{ds} = 0.1$ V with the parameters values of Table 1

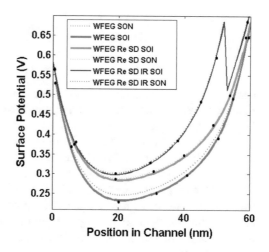

Fig. 3 Surface potential distribution along the channel position of WFEG Re S/D IRSOI/SON MOSFETs for different values of V_{ds}(0.1, 0.5, 0.75 V) and $V_{GS} = 0.1$ V

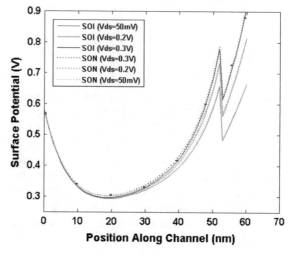

Fig. 4 Variation of surface
potential for different
channel length, all other
parameters value listed in
Table 1

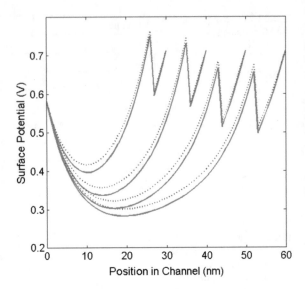

against drain bias variation or drain-induced barrier lowering (DIBL) effect. The
figure also reveals that SON structure poses relatively higher immunity against this
V_{ds} variation compared to its SOI counter parts.

Continuous reduction in the device dimension increases the charge sharing effect
which results in gradual shifts of potential minima toward the source side. Although
from Fig. 4, it is established that the proposed structure reinstates the symmetry
of the potential profile, revealing its immunity against DIBL and other SCEs. Further, the presence of IR region reduces the potential at the drain side, thereby sup-
pressing the effects of HCEs. Again from Fig. 4, it is evident that SON structure
provides slight higher potential minima as compared to SOI and reveals its lower
parasitic effects.

4   Conclusion

In this paper, a unique structure of work function engineered gate (WFEG) recessed
S/D SOI MOSFETs with high-k dielectric region in between channel and drain
end has been developed. Due to the combined benefits of both WFEG recessed S/D
structure and high-k dielectric material in IR region, the proposed structure confirmed
a better device performance in terms drain-induced barrier lowering (DIBL) and
hot-carrier effects (HCE). Comparative performance analysis of SOI and SON in
terms of surface potential distribution reveals that SON has better immunity against
various SCEs than its counterpart. Therefore, the proposed recessed S/D IR SOI/SON
structure may be used in the nanometer regime to optimize the desired performance
of the device parameter.

References

1. Miura-Mattausch, M., Mattausch, H.J., Ezaki, T.: The Physics and modeling of MOSFET. World Scientific Publishing Co. Pte. Ltd., Singapore (2008)
2. Colinge, J.P.: Silicon on Insulator Technology: Materials to VLSI, 2nd edn. Kluwer Academic Publishers, Norwell (1997)
3. Colinge, J.P.: Multi-gate SOI MOSFETs. Microelectron. Eng. **84**, 2071–2076 (2007)
4. Youssef Hammad, M., Schroder, D.K.: Analytical modeling of the partially-depleted SOI MOS-FET. IEEE Trans. Electron Dev. **48**(2), 252 (2001)
5. Chen, J., Luo, J., Wu, Q., Chai, Z., Yu, T., Dong, Y., Wang, X.: A Tunnel diode body contact structure to suppress the floating-body effect in partially depleted SOI MOSFETs. IEEE Electron Dev. Lett. **32**(10) (2011)
6. Deb, S., et al.: Two dimensional analytical-model-based comparative threshold performance analysis of SOI-SON MOSFETs. J. Semiconductor **32**(10) (2011)
7. Deb, S., Basanta Singh, N., Das, D., De, A.K., Sarkar, S.K.: Analytical I-V model of SOI and SON MOSFETs: a comparative analysis. Int. J. Electron. **98**(11), 1465–1481 (2011)
8. Manna, B., Sarkhel, S., Ghosh, A., Singh, S.S., Sarkar, S.K.: Dual material gate nanoscale SON MOSFET: for better performance. Int. J. Comput. Appl. (IJCA) (2013). ISBN: 973-93-80875-27-15
9. Deb, S., et. al.: Work function engineering with linearly graded binary metal alloy gate electrode for short channel SOI MOSFET. IEEE Trans. Nanotechnol. **11**(3), 472–478
10. Manna, B., Sarkhel, S., Islam, N., Sarkar, S., Sarkar, S. K.: Spatial composition grading of binary metal alloy gate electrode for short-channel SOI/SON MOSFET application. IEEE Trans. Electron. Dev. **59**(12), 3280–3287 (2012)
11. Zhang, Z., Zhang, S., Chan, M.: Self-align recessed source/drain ultrathin body SOI MOSFET. IEEE Electron. Dev. Lett. **25**, 740–742 (2004)
12. Reddy, V., Jagadesh Kumar, M.: A new dual-material double-gate (DMDG) nanoscale SOI MOSFET-twodimensional analytical modeling and simulation. IEEE Trans. Electron. Dev. **4**(2), 260–268 (2005)
13. Young, K.K.: Short-channel effects in fully depleted SOI MOSFET's. IEEE Trans. Electron. Dev. **36**, 399–402 (1989)
14. Svilicic, B., et al.: Analytical models of front- and back-gate potential distribution and threshold voltage for recessed source/drain UTB SOI MOSFETs. Solid State Electron. **53**, 540–547 (2009)

Network Security

A Mobile User Authentication Technique in Global Mobility Network

Sudip Kumar Palit and Mohuya Chakraborty

Abstract Anonymous user authentication always is a paramount job in global mobility network (GLOMONET). In GLOMONET, a mobile user can move from one place to another place causes changes of mobile network from one to another. As authentication server can authenticate only its registered users, it cannot verify other mobile users. Furthermore, in global mobility network, communication channel is public. An adversary of the network can get access of all transmitted messages over the channel. In such a situation, several network security attacks can be initiated by an adversary to decrypt the messages. Therefore, mobile users as well as network provider's information may come in risk. A robust anonymous user authentication and key agreement technique required to ensure the protection of such information. Several authentication protocols have been designed in GLOMONET in recent years. Unfortunately, most of them are unprotected against various network security attacks. Therefore, in this paper, we proposed a rigid authentication protocol in GLOMONET to overcome all the deficiencies of the previous work. Performance analysis of our protocol shows that it is reliable and even more effective compared to other existing protocols.

Keywords Global mobility network · Authentication · Network security · AVISPA

1 Introduction

GLOMONET makes roaming service be available at everywhere. Mobile users can get the network service from the home network. But whenever they transit from the coverage of the home network, they can avail the same service from any available foreign network, with the condition that the foreign network has roaming agreement with the home network. However, the mobile users do register themselves under the

S. K. Palit (✉)
University of Engineering and Management, Kolkata, India
e-mail: palit.sudipkumar@gmail.com; palit_sudip_kumar@yahoo.co.in

M. Chakraborty
Institute of Engineering and Management, Kolkata, India

© Springer Nature Singapore Pte Ltd. 2020
M. Chakraborty et al. (eds.), *Proceedings of International Ethical Hacking Conference 2019*, Advances in Intelligent Systems and Computing 1065, https://doi.org/10.1007/978-981-15-0361-0_14

home network. Therefore, foreign network has no authentication information about the mobile users who want roaming service in its network area. In this scenario, foreign network depends on the home network of such mobile users for the authentication purpose. As the communication takes place in the network is in open channel, anyone if want can get access of the transmitted messages. Therefore, different network security attacks, including replay attack, man in middle attack, impersonation attack, etc. may possible in GLOMONET to break the solidness of the network.

Various user authentication protocols [1–22] for GLOMONET have been designed over the last few years. The securities of these protocols were based on symmetric encryption/decryption function, secure hash function, elliptic curve cryptography (ECC), modular exponent operation, etc. The asymmetric key operations like ECC, modular exponent operation takes more time than secure hash function and symmetric key operations like encryption/decryption function. As the mobile phone always does not contain powerful processor, the cryptographic operation which takes less time is more suitable. However, the protocols [1–22] designed in the GLOMONET; some of these have some deficiencies and cannot resist certain kinds of network security attacks. Therefore, a powerful protocol required which can prevent all known network security attacks as well as suitable for low power processor of mobile device. In this paper, we have proposed a powerful protocol which can not only resist all kind of known attacks but also take less processing time compare to other protocols of recent time.

Our contribution to this paper can be described as follows. Literature survey of different research journals in GLOMONET has performed at first. This segment of our paper described the deficiencies of the previous papers. Then we formulate our own protocol where we have eliminated the specified deficiencies of the previous papers. Thereafter, the soundness of our protocol has been proved with the help of formal and informal security analysis. Finally, we compare our protocol with some other protocols of same topic. The organization of the paper is as follows. Section 2 contains literature survey of some papers [1–22]. Section 3 describes our proposed protocol. Section 4 portrays formal and informal security exploration. A relative performance analysis of different protocols is depicted in Sect. 5. Finally, Sect. 6 concludes our paper.

2 Literature Survey

In 2008, Wu et al. [1] designed a protocol on anonymity in wireless network. They used symmetric encryption, decryption function in their proposed protocol. However, in 2009, Zeng et al. [2] showed that the protocol of Wu et al. [1] does not withstand user anonymity attack. Again, in 2009, Chang et al. [3] proposed an authentication scheme in GLOMONET where they used simple hash function and exclusive OR function to reduce operating cost for mobile users.

However, in 2011, Zhou and Xu suggested a protocol [4] based on the Decisional Diffie–Hellman (DDH) notion. Anyway, due to uses of exponential function, the protocol took higher computational time and cost. In 2012, Mun et al. [5] proved Wu et al. [1] protocol fails to resist user anonymity attack and also fails to assure forward secrecy. Then they presented an authentication protocol where they apply Elliptic Curve Diffie–Hellman[ECDH] notion to solve the pitfalls of Wu et al. [1] protocol.

However, in 2012, Hsieh and Leu [6] designed a scheme depend on ECDH notion. Again in the same year, Kim and Kwak [7] showed that Mun et al. [5] protocol has weakness against man in middle attack and replay attack. In 2013, Jiang et al. [8] suggested an enhanced authentication scheme in GLOMONET. However, in the same year, Lee [9] showed that Chang et al. [3] schemes do not resist against impersonation attack and contravene the session key security. In his proposed scheme, Lee [9] used a temporary identity of mobile user MU so that identification of MU is not revealed to the adversary during dispatch of messages in open channel. Again, in 2013, He et al. [10] showed that the protocol proposed by Hsieh and Leu [6] has weakness against user anonymity attack. Afterward, they offered a scheme on ECDH problem to ensure security against user anonymity attack. In 2013, Wen et al. [11] proved that the scheme offered by Jiang et al. [8] is not secure against replay attack, stolen verifier attacks, and denial-of-service attacks.

In 2013, an authentication protocol was designed by Zhao et al. [12] on GLOMONET. They showed that Mun's [5] protocol is not only unsafe against offline password guessing attack, impersonation attack, insider attack but also was not able to produce user's anonymity and proper mutual authentication. They designed a new protocol based on ECDH notion where they overpower the weakness of Mun's [5] protocol. In 2014, Kuo et al. [13] designed a protocol based on the authorization of mobile user in GLOMONET using ECDH notion. However, in 2015, Gope and Hwang [14] proposed a protocol where they eliminate the security deficiencies of Wen et al. [11] like offline guessing attack, forgery attack, and unfair key agreement. In this protocol, authors used modular operation, symmetric encryption, decryption operation. Furthermore in 2016, Gope and Hwang [15] proved that the scheme offered by Zhou and Xu [4] has weakness in replay attack, insider attack, and forgery attack. Then they proposed a lightweight authentication protocol in GLOMONET. In this protocol, authors used simple one-way hash function to overcome the deficiencies of Zhou and Xu's [4] scheme. As execution time of hash function is much less compare to symmetric and asymmetric key operation, the protocol they [15] proposed took immensely fewer time compare to other protocols [4, 5, 8]. In 2016, Karuppiah et al. [16] proposed a lightweight authentication scheme. In the same year, Gope and Hwang [17] proposed a protocol of mutual authentication and key agreement in global mobility network. They used symmetric key operation and one-way hash function in the protocol. Again in 2016, Xu and Wu [18] introduced a three-factor user authentication key agreement protocol in GLOMONET. Although, session key renewal phase has not designed in this scheme. However, in 2017, Arshad and Rasoolzadegan [19] proved the protocol of Karuppiah et al. [16] fails to resist offline password guessing attack and it also does not provide perfect forward secrecy property. In the same year, Madhusudhan and Suvidha [20] showed that Gope and Hwang's [17] scheme is vulnerable to stolen smart card attack, replay attack, forgery attack, and offline password guessing attack. Moreover, their scheme does not maintain user anonymity. To overcome the security inadequacy, they proposed another user authentication scheme which can preserve the user anonymity property in GLOMONETs.

Later, in 2017, Lee et al. [21] designed a scheme where they proved Mun et al. [5] protocol has weakness against masquerade attack, user impersonation attack, man in the middle attack and does not provide perfect forward secrecy. Then they proposed a scheme using simple hash function and exclusive OR function to overcome the flaws of Mun et al. [5] but in Nov 2017, Park et al. [22] proved that the protocol presented by Lee et al. [21] is unsafe against User impersonation attack, offline password guessing attack and flaws in proper mutual authentication, perfect forward secrecy. Then they [22] proposed a scheme in GLOMONET where they eliminated the deficiencies of Lee et al.'s [21] scheme.

3 The Proposed Protocol

Our protocol consists of three entities namely MU, FA, HA, and five phases. The phases are 1. Mobile user registration phase, 2. Foreign agent registration phase, 3. Mutual authentication and session key agreement phase, 4. Session key renewal phase, 5. Password altered phase. Table 1 described the different notations used in our protocol. The descriptions of phases are given below one by one.

Table 1 Symbol table

Notation	Description
MU	Mobile user
FA	Foreign agent
HA	Home agent
IDx	Identity of an entity x
m, m1, Nm	Random numbers generated by MU
F, Nf, Nf2	Random numbers generated by FA
Nh	Random numbers generated by HA
Ki, Ski, SKha	Secret values of HA
SKmf	Shared secret session key between MU and FA
H()	One-way hash function
x‖y	Concatenation operation between x and y
$x \oplus y$	Exclusive OR operation between x and y

3.1   Mobile User Registration Phase

In this phase, a new mobile user at first registered under a suitable mobile network (home agent). The steps are as follows.

Step1. MU chooses an identity IDmu, a password PWmu and a nonce m. Computes PIDmu = H(IDmu‖m), MPW = H(PWmu‖IDmu)
Then send <PIDmu> to HA in a secure channel.
Step2. HA maintain a small database of triplet [(Si,Ki,Ski), $1 <= i <= n$: n is a small number, may be taken as one unit for each 1000 subscriber] in its server in a very secure manner. After receiving PIDmu, HA randomly chooses a number i from 1 to n. Then get Si, Ki, SKi corresponding to i from the secret table.
Computes AMU = H(PIDmu‖H(Ki)), EMU = H(Ski‖H(PIDmu)).
Then sends <AMU, EMU, Si> to MU in a secure channel.
Step3. After receiving <AMU, EMU, Si> MU computes
BMU = Si⊕ H(IDmu‖PWmu), LMU = H(MPW‖Si),
CMU = (AMU + EMU) ⊕ H(IDmu ‖ MPW)
Then stores <PIDmu,BMU,LMU,CMU> in the smartcard.

3.2   Foreign Agent Registration Phase

In this phase, a mobile network (foreign agent) at first register under any other mobile network (home agent). The steps are as follows

Step1. FA chooses an identity IDfa and a nonce f.
Computes Kfa = H(IDfa ‖ f)
Then sends <IDFa,Kfa> to HA in a secure channel.
Step2. After receiving <IDfa,Kfa> from FA, HA computes
SKfa = H(IDfa‖SKha).
Then stores <Kfa,SKfa> in it's database and sends <SKfa> to FA in a secure channel.
Step3. After receiving <SKfa> rom HA, FA stores <Kfa,SKfa> in its database.

3.3   Mutual Authentication and Session Key Agreement Phase

In this phase, a registered mobile user MU can avail network service from a foreign agent FA. At first, MU sends authentication request to FA. After receiving request from MU, FA does communicate with respective HA. If the MU is genuine, then a session key is generated between MU and FA. The steps are as follows.

Step1. MU Inputs IDmu, PWmu and computes
MPW = H(PWmu‖IDmu), Si = BMU ⊕ H(IDmu‖PWmu),

Then Checks LMU ? = H(MPW$\|$Si). If it is true then MU is genuine and then MU computes (AMU + EMU) = CMU $\oplus$ H(IDmu$\|$ MPW)

Then MU Select two random nonces m1, Nm.

Computes PIDnew = H(IDmu$\|$m1), DMU = (Nm$\|$PIDnew)$\oplus$ H(AMU$\|$Si)

Qm = H(Nm$\|$EMU), FMU = EMU $\oplus$ H(AMU$\|$Nm), FMS = FMU $\oplus$ Si. Then sends MSG1 = <DMU,Qm,IDha,PIDmu,FMU,FMS> to FA in a public channel.

Step2. After receiving <DMU,Qm,IDha,PIDmu,FMU,FMS> from MU, FA Chooses a random number Nf, then computes Qf = H(Qm $\|$Nf$\|$SKfa), AFA = H(SKfa$\|$Kfa)$\oplus$ Nf.

Finally sends <MSG,Qf,AFA,IDfa> to the respective HA.

Step3. After receiving <MSG,Qf,AFA,IDfa> from FA, HA Computes

Si = FMU $\oplus$ FMS, find Ki, SKi corresponding to Si, then

Compute AMU = H(PIDmu$\|$H(Ki)), EMU = H(SKi$\|$H(PIDmu))

SKfa = H(IDfa$\|$SKha), find Kfa,

Compute Nf = AFA $\oplus$ H(SKfa$\|$Kfa), (Nm$\|$PIDnew) = DMU $\oplus$ H(AMU$\|$Si)

EMU' = FMU $\oplus$ H(AMU$\|$Nm).

Then checks EMU ? = EMU' and Qf ? = H(Qm$\|$Nf$\|$SKfa), if so, then HA authenticates MU and FA.

After that, HA Computes AMUnew = H(PIDnew$\|$H(Ki)), EMUnew = H(SKi$\|$PIDnew)

AMU1 = AMUnew $\oplus$ AMU, EMU1 = EMUnew $\oplus$ AMU

Then, generate a random nonce Nh, compute

AHA = (AMU1$\|$EMU1$\|$EMU$\|$Nm)$\oplus$ H(SKfa$\|$Kfa$\|$Nh).

Finally sends <AHA,Nh> to FA in a public channel.

Step4. After receiving <AHA,Nh> from HA, FA Retrieves

(AMU1$\|$EMU1$\|$EMU$\|$Nm) = AHA $\oplus$ H(SKfa$\|$Kfa$\|$Nh)

Then checks Qm ? = H(Nm$\|$EMU), if so then FA authenticates HA and MU.

Chooses a random number Nf2, compute

BFA = (AMU1$\|$EMU1$\|$Nf2)$\oplus$ H(EMU$\|$Nm),

Qf1 = H(DMU$\|$AMU1$\|$Nf2)

Finally sends <BFA,Qf1> to MU.

Step5. After receiving <BFA,Qf1> from FA, MU retrieves

(AMU1$\|$EMU1$\|$Nf2) = BFA $\oplus$ H(EMU$\|$Nm)

Checks Qf1 ? = H(DMU$\|$AMU1$\|$Nf2). If so then MU authenticates FA and HA.

Compute AMUnew = AMU1 $\oplus$ AMU, EMUnew = EMU1 $\oplus$ AMU

CMUnew = (AMUnew + EMUnew)$\oplus$ H(IDmu$\|$MPW)

Then replace CMU,PIDmu with CMUnew,PIDnew in the smart card.

Compute session key SKmf = H(Nm$\|$Nf2$\|$EMU)

Store the session key for future reference.

Also compute Qm1 = H(Nm$\|$Nf2)

Finally sends <Qm1> to FA.

Step6. After receiving <Qm1> from MU, FA

Checks Qm1 ? = H(Nm$\|$Nf2). If so then compute session key

SKmf = H(Nm$\|$Nf2$\|$EMU)

Store the session key for future reference.

3.4 Session Key Renewal Phase

In this phase, MU and FA can renew their session if they want and they do this without any interaction with HA. The steps are as follows.

Step1. MU chooses new random number Nm1 and computes
$FMU = Nm1 \oplus H(Nm\|Nf2\|EMU)$, $Qm" = H(Nm1\|EMU)$.
Then MU sends <FMU, Qm"> to FA in a public channel.
Step2. After receiving <FMU, Qm">, FA Computes
$Nm1 = FMU \oplus H(Nm\|Nf2\|EMU)$
Checks $Qm"$? $= H(Nm1\|EMU)$, if so then chooses new random nonce Nf1, computes $CFA = Nf1 \oplus H(Nm\|Nf2\|EMU)$, $Qf" = H(Nf1\|EMU)$.
Finally sends <CFA,Qf"> to MU in public channel.
Step3. After receiving <CFA,Qf"> from FA, MU computes
$Nf1 = Qf" \oplus H(Nm\|Nf2\|EMU)$
Checks $Qf"$? $= H(Nf1 \oplus EMU)$, if so then update new session key
$SKmf' = H(Nm1\|Nf1)$, $Qm2 = H(Nm1\|Nf1\|EMU\|EMU')$.
Finally sends <Qm2> to FA in public channel.
Step4. After receiving <Qm2> from MU, FA
Checks $Qm2$? $= H(Nm1\|Nf1\|EMU\|EMU')$, if so then
Update session key $SKmf' = H(Nm1\|Nf1)$

3.5 Password Altered Phase

In this phase, MU can change his/her password without any interaction with HA. The steps are as follows.

MU inputs IDmu, PWmu, compute $MPW = H(PWmu\|IDmu)$, $Si = BMU \oplus H(IDmu\|PWmu)$
Checks LMU? $= H(MPW\|Si)$, if so then compute
$(AMU + EMU) = CMU \oplus H(IDmu \| MPW)$, choose a new password PWmu', compute $MPWnew = H(PWmu'\|IDmu)$, $CMUnew = (AMU + EMU) \oplus H(IDmu\|MPWnew)$
$BMUnew = Si \oplus H(IDmu\|PWmu')$
Finally replace BMU,CMU with BMUnew, CMUnew in the smart card.

4   Formal and Informal Security Analysis

4.1   Informal

In this section, we illustrate how our protocol can resist different network security attacks.

User Anonymity and Untraceability
The communication takes place between MU, HA, and FA in the public channel all are change every time hence the communications are anonymous and untraceable. For example, MU sends (DMU,Qm,IDha,PIDmu,FMU,FMS) to FA during mutual authentication and session key agreement phase. MU's identity IDmu is included in PIDmu where PIDmu = H(IDmu∥m). Here, m is a random number which changes at every mutual authentication and session key agreement phase. Therefore, PIDmu also changes in every phase making identity of MU anonymous. All other parameters like DMU, Qm, FMU, FMS are computed with the help of random numbers Nm, m, m1. This dynamic structure of the parameters made them anonymous and untraceable. All the parameters in our protocol transferred between MU, HA, FA are formed in the same way. Therefore, all communication takes place are anonymous and untraceable. Furthermore, the real identity IDmu of MU keeps concealed even to HA. This feature helps to protect the identities of mobile user from being misused.

Perfect Mutual Authentication
In this protocol, all the entities mutually authenticate each other in pair as described below.

HA confirms the authenticity of MU by examining EMU' ? = EMU. Here, EMU = H(SKi ∥ H(PIDmu)), EMU' = FMU ⊕ H(AMU∥Nm) and FMU = EMU ⊕ H(AMU∥Nm). HA computes EMU with its secret value SKi. AMU also calculated by HA with its secret value Ki. Besides HA get FMU, Nm from MU via FA. Therefore, if the above condition became true, it would be possible if and only if valid MU and HA compute these parameters. Therefore, the above relation shows that there exists a mutual authentication between MU and HA.

Similarly, the mutual authentication between FA and HA can be verified by the relation Qf ? = H(Qm∥Nf∥SKfa). Here, Qf and Qm received by HA from FA. SKfa is a secret key shared between FA and HA. Therefore, above relation hold well for genuine HA, FA pair.

On the other hand, FA can verify the genuineness of MU and HA by verifying Qm ? = H(Nm∥EMU). Here, FA obtained Qm from MU whereas Nm, EMU are obtained by decrypting AHA by computing (AMU1∥EMU1∥EMU∥Nm) = AHA ⊕ H(SKfa∥ Kfa∥Nh). The key SKfa is shared between FA and HA. As AHA received from HA, so if the relation is true, then FA can verify the genuineness for both HA and MU. Therefore, mutual authentication between FA and MU, FA and HA also preserved in our scheme.

Lastly, MU verifies the genuineness of FA and HA by verifying $Qf1 = H(DMU ||AMU1||Nf2)$. Here, $AMU1 = AMUnew \oplus AMU$. Here, AMU is only known to HA and MU. Again Nf2 is obtained by decrypting BFA by computing $(AMU1||EMU1||Nf2) = BFA \oplus H(EMU||Nm)$. Here, EMU is shared secret between MU and FA. Therefore, if the above relation is true, then we can say MU,FA pair and MU, HA pair mutually authenticate each other.

As every participant mutually authenticates rest entities, we can say that our scheme follows a perfect mutual authentication.

Replay Attack

In our protocol, if an intruder Æ accumulates all the messages that have been transmitted in the earlier phases and wants to execute a replay attack by sending the same message to extract information from the system, it will be unsuccessful because all the parameters of the said messages are computed with random nonces m, m1, Nm, Nf, Nf2, Nh which are altered in every phase. Therefore, repeating messages is not feasible which obstructs Æ to perform a replay attack.

Man in Middle Attack

In our protocol, there are five messages are transmitted during mutual authentication and session key agreement phase. First, MU send message to FA. Second, FA send message to HA. Third, HA send message to FA. Fourth, FA send message to MU. Fifth, MU send message to FA. Among these five cases, no man in middle attack possible in second to fifth. Because in second case, each of FA and HA can check the genuineness of other by their secret key Kfa and SKfa. Similarly in third case, the genuineness is being checked by the secret key Kfa, SKfa. So, no intruder from middle can break the security of the protocol without knowing the secret keys. In fourth and fifth cases, FA and MU authenticate each other by the shared secret EMU between them. Without knowing the value of EMU, no intruder can break the secrecy of our protocol. Although there is no way to know the value of EMU as it is transmitted in encrypted form with the key values AMU, SKfa. Here, AMU is shared secret between MU and HA, whereas SKfa is a shared secret between FA and HA. Only possibility of man in middle attack exists in first case as there is no way to check the genuineness of an unknown MU for an FA. But this possibility will loss as HA can check the legitimacy of the attacker by the secret shared key AMU between MU and HA. Therefore, no man in middle attack is possible in our protocol.

Stolen Smartcard Attack

If Æ can anyway manage to get or stole smartcard of MU, then he/she has to input identity and password of the legitimate user which is not possible in normal cases. Thus, local password verification of MU obstructs Æ from illegal access.

Insider Attack

In our protocol, the identity IDmu of a mobile user MU is encrypted with a nonce M by one-way hash function to generate PIDmu. Therefore, an insider in HA cannot obtain real identity IDmu of MU from PIDmu. Again, as we have seen in stolen smartcard attack that MU can access his/her smartcard after unlocking it with his/her identity and password. Therefore, no insider in-home agent can get an advantage from our protocol without knowing this personal information. Again, no insider in FA can get the real identity of MU. The useful information that an FA can get is EMU for the current session. But this value also changed for subsequent mutual authentication and session key agreement phase. Therefore, no insider in FA can get advantage from our protocol. Lastly, an insider may come from MU's side by registered his/herself under a home agent and may try to break the system. In that case, the insider will has (PIDmu,BMU,LMU,CMU) along with identity IDmu and password PWmu of his/her. Then the insider may try to extract information about the secret key Ski, Ki or Si of HA or try to manage identity or password of another valid mobile user from these stored values in smartcard. Anyway, if he/she can manage to extract AMU, EMU, Si from the stored data in the smart card, then also it is not possible to manage the secrets of HA and other MUs. Here, AMU = H(PIDmu ∥ H(Ki)), EMU = H(SKi ∥ H(PIDmu)). From this relation, if insider knows the value of PIDmu, then it is not possible to extract value of Ki and SKi. Again as each MU has a different PIDmu, the AMU and EMU values are also different for each MU. Therefore, AMU, EMU pair of particular insider cannot help him/her to extract information about other MU. Therefore, in our protocol, no insider attack is possible.

Impersonation Attack

If an adversary wants to impersonate MU, then he/she has to send a valid message MSG = (DMU,Qm,IDha,PIDmu,FMU,FMS) to FA. The parameters DMU, Qm, FMU, FMS are encrypted with either by AMU or by EMU. PIDmu is encrypted by IDmu. Now (AMU + EMU) = CMU ⊕ H(IDmu ∥ MPW), MPW = H(PWmu∥IDmu). The adversary may collect the smart card information PIDmu,BMU,LMU,CMU of the targeted MU. Now to extract AMU and EMU, the adversary has to produce IDmu and MPW of the victim MU. But as the identity IDmu or password PWmu of that MU is not known to the adversary. Therefore, mobile user impersonation is not possible in our scheme.

Again an adversary can impersonate FA by sending the message (MSG,Qf,AFA,IDfa) to HA. Here, AFA = H(SKfa∥Kfa)⊕ Nf, Qf = H(Qm ∥ Nf ∥ SKfa). Therefore, to impersonate as genuine FA, the adversary has to produce the original value of SKfa, Kfa. As these values are secret between FA and HA, impersonation as FA is not possible in our scheme. There is another possibility of impersonate as FA when FA sends (BFA,Qf1) to MU. Here, BFA = (AMU1∥EMU1∥Nf2) ⊕ H(EMU∥Nm), Qf1 = H(DMU∥AMU1∥Nf2). Further AMU1 = AMUnew ⊕ AMU, EMU1 = EMUnew ⊕ AMU. Here, all data other than Nf2 is precalculated. To cheat MU, the adversary has to produce a valid value for AMU, EMU to compute these parameters. As these values are known only to MU and HA, the forgery of the adversary will be identified in the MU side.

Lastly, adversary can impersonate as MU by sending (AHA,Nh) to FA. Here, AHA = (AMU1∥EMU1∥EMU∥Nm) ⊕ H(SKfa∥Kfa∥Nh). As the values of SKfa, Kfa are secret, valid message cannot be produced by the adversary in that case too. Therefore, an impersonation attack is not possible in our protocol.

Offline Password Guessing Attack

In the suggested protocol, if the smartcard of MU is stolen, then extracting the loaded information in the smart card is possible by power analysis attack [23, 24]. Hence, MU's personal information is no safer due to power analysis attack. In case of stolen smartcard, if intruder Æ is somehow managed to take out necessary information PIDmu,BMU,LMU,CMU from smartcard, he/she has to proceed local password verification of MU where Æ has to input PWmu, IDmu of genuine MU. To predict correctly, idMU and pwMU is very difficult. Therefore, our protocol gives dependability against offline password guessing attack.

Perfect forward Secrecy

Let adversary Æ in some way has managed to collect all the communication between MU, FA, and HA of the former sessions. He/she also by any way compromised SKha the long-term key of HA. Therefore, Æ can calculate SKfa = H(IDfa‖SKha) where IDfa is available from previous message. Although SKfa is available for Æ, another key attribute Kfa is not available to Æ. Therefore, the session key SKmf = H(Nm‖Nf2‖EMU) between FA and MU could not be calculated by the Æ because none of the information to compute session key is available for the adversary Æ. Therefore, our protocol provides perfect forward secrecy.

Local Password Verification

Our protocol provides safety against illegitimate access of MU's smartcard by obstructing Æ in the local password verification. If the intruder Æ has somehow managed smartcard of other MU, he/she has to pass local password verification phase by inputting identity IDmu and password PWmu of the legitimate user. This information is not available to Æ. This feature gives safety against unauthorized access.

4.2 AVISPA Simulation Tool for Formal Security Analysis

In this segment, we have conducted a formal security verification of our protocol with the help of widely familiar Automated Validation of Internet Security Protocols and Applications (AVISPA) simulation tool [25]. This simulation tool has a great number of uses as formal security verification to justify whether the authentication schemes are stable against replay attack and man in middle attack. AVISPA implements a role-based language, which is known as high-level protocol simulation language (HLPSL) [26]. The translator of HLPSL, HLPSLIF translates the hlpsl source language to intermediate format (IF). Ultimately, the intermediate format converted to the output format (OF) with the help of any of the four back-ends tools: On-the-fly Model-Checker (OFMC), the CL-based Attack Searcher (CLAtSe), SAT-based Model-Checker (SATMC), or Tree Automata-based Protocol Analyzer (TA4SP).

HLPSL specification:

As per the notation of AVISPA, our protocol has three basic roles: mobileuser means Mobile User (MU), foreignagent means Foreign Agent (FA), and homeagent means Home Agent (HA). Besides these roles, two more roles session and environment represent the session and environment of our source code. The roles of MU, HA, and FA have demonstrated in Figs. 1, 2, and 3, respectively. The roles of session and environment have demonstrated in Fig. 4. Here, environment role describes the secrecy and authentication goal of our protocol to verify whether our scheme is secure or not. The role of MU can be described in the following way:

When MU receives the start signal, MU moves from state 0 to 3. Thereafter, MU sends the registration request PIDmu to HA via a secure channel. During this time, MU retains PIDmu as a secret shared between MU and HA whereas retains

```
%%%%% Role for MU %%%%%
role mobileuser(MU,HA,FA:agent,  SKmuha:symmetric_key,
          H:hash_func,SND,RCV:channel(dy))
% player: the mobile user MU
played_by MU
def=
local State : nat,M,M1,IDmu,PWmu,PIDmu,IDha,Nf2,Qm1,Ki,SKi,Nm,PIDnew,Qm,
          FMU,DMU,FMS,Si:text
          const sp1,sp2,sp3,mu_fa_nm,mu_fa_m1,fa_mu_nf2:protocol_id
init State:=0
transition
% Mobile user registration phase
1.State=0∧RCV(start)=|>
State':= 3∧ M':=new() ∧ PIDmu':= H(IDmu.M')
          % Send registration request <PIDmu> to HA securely
          ∧SND({PIDmu'}_SKmuha)
          ∧secret({PIDmu'},sp1,{MU,HA})∧secret({IDmu,PWmu,M'},sp2,{MU})
          %Receive registration reply<AMU,EMU,Si> from HA securely
2.State=3∧RCV({H(H(IDmu.M).H(Ki)).H(SKi. H(IDmu.M)).Si}_SKmuha)=|>
          State':=5∧secret({Ki,SKi},sp3,{HA})
%authentication & establishment of session key phase
          ∧ M1':=new() ∧ Nm' :=new()∧PIDnew':=H(IDmu.M1')
          ∧Qm':= H(Nm'.H(SKi. H(IDmu.M)))∧FMU':=xor(H(SKi. H(IDmu.M)),
          H(H(H(IDmu.M).H(Ki)).Nm'))∧FMS':=xor(xor(H(SKi. H(IDmu.M)),
          H(H(H(IDmu.M).H(Ki)).Nm')),Si)∧DMU':=xor((Nm'.H(IDmu.M1')),
          (H(H(H(IDmu.M).H(Ki)).Si)))
% Send message {DMU,Qm,idHA,pidMU,FMU,FMS} to FA via open channel
          ∧SND(DMU'.Qm'.IDha.PIDmu.FMU'.FMS')
%MU has freshly generated random numbers Nm and snew for FA
          ∧witness(MU,FA,mu_fa_nm,Nm')∧witness(MU,FA,mu_fa_m1,M1')
% Receive message<BFA,Qf1,Nf2> from FA via open channel
3. State=5∧RCV(xor((xor(H(H(IDmu.M1).H(Ki)), H(H(IDmu.M).H(Ki))).
          xor(H(SKi. H(IDmu.M1)), H(H(IDmu.M).H(Ki)))),H(H(SKi.
          H(IDmu.M)).Nf2')).H(xor((Nm. H(IDmu.M1)),(H(H(H(IDmu.M).
          H(Ki)).Si))). xor(H(H(IDmu.M1).H(Ki)), H(H(IDmu.M).H(Ki))).
          Nf2').Nf2')=|>
State':=9∧Qm1':=H(Nm.Nf2')
%Send message <Qm1> to FA via open channel,
          ∧SND(Qm1')∧request(FA,MU,fa_mu_nf2,Nf2')
end role
```

Fig. 1 Role specification for mobile user

```
%%%%% Role for HA %%%%%
role homeagent(MU, HA, FA:agent,
 SKmuha:symmetric_key,H:hash_func,SND,RCV:channel(dy))
%player the home agent HA
played_by HA
def=local State:nat,
           IDmu,M,PIDmu,PWmu,AMU,EMU,Ki,SKi,Si,SKha,Nm,M1,IDha,Nh,AHA,
           Nf,SKfa,IDfa,Kfa:text
           const sp1, sp2, sp3, sp4, ha_fa_nh,fa_ha_nf: protocol_id
init State:=1
transition
%Mobile User Registartion Phase
%Receive registration request <PIDmu> to from MU securely
1. State=1∧RCV({H(IDmu.M')}_SKmuha)=|>
   State':=4∧secret({PIDmu},sp1,{MU,HA})∧secret({IDmu,PWmu,M},sp2,{MU})
           ∧AMU':=H(H(IDmu.M').H(Ki))
           ∧EMU':=H(SKi.H(IDmu.M'))
      ∧secret({Ki,SKi}, sp3, {HA})
% Send registartion reply to FA secretly
           ∧SND({AMU'.EMU'.Si}_SKmuha)
%authentication and establishment of session key phase
%Receive message {MSG,Qf,AFA,idFA} from FA via public channel
2.State=4∧RCV(xor((Nm'. H(IDmu.M1')),(H(H(H(IDmu.M).H(Ki)).Si))).H(Nm'.
           H(SKi. H(IDmu.M))).IDha.H(IDmu.M).xor(H(SKi. H(IDmu.M)),
           H(H(H(IDmu.M).H(Ki)).Nm')).xor(xor(H(SKi. H(IDmu.M)),H(H(H(IDmu.M).
           H(Ki)).Nm)),Si).H(H(Nm'.H(SKi. H(IDmu.M))).Nf.SKfa).
           xor(H(SKfa.Kfa),Nf').IDfa)=|>
State':= 7∧secret({Kfa,SKfa},sp4,{FA,HA})∧Nh':=new()∧AHA':=
           xor((xor(H(H(IDmu.M1).H(Ki)), H(H(IDmu.M).H(Ki))).xor(H(SKi.
           H(IDmu.M1)), H(H(IDmu.M).H(Ki))).H(SKi. H(IDmu.M)).Nm'),
           H(SKfa.Kfa.Nh'))
% send message {AHA, Nh} to FA via open channel
           ∧SND(AHA'.Nh')
           ∧witness(HA,FA,ha_fa_nh,Nh')
%HA's acceptance of the value Nf generated for HA by FA
           ∧request(FA,HA,fa_ha_nf,Nf')
end role
```

Fig. 2 Role specification for home agent

IDmu,PWmu and a randomly generated nonce M as secret to himself. After that, MU accepts registration reply from HA and then changes the state from 3 to 5 during mutual authentication and session key agreement phase. Thereafter, MU despatches the message DMU,Qm,idHA,pidMU,FMU,FMS to FA via public channel. During this time, MU announces witness of the event that MU has generated a fresh random numbers Nm and M1 for FA.

Thereafter, MU accepts the message BFA, Qf1, Nf2 from FA in a public channel and moves from state from 5 to 9. Lastly, MU despatches Qm1 to FA through a public channel. Likewise, the role specification of HA and FA is also demonstrated in Figs. 2 and 3, respectively.

Analysis of AVISPA simulation result:

Here, we described the results of the AVISPA analysis using On-the-fly Model-Checker (OFMC) and the CL-based Attack Searcher (CLAtSe) back ends to confirm

```
%%%%% Role for FA %%%%%
role foreignagent(MU, HA, FA:agent,H:hash_func,SND,RCV:channel(dy))
% player: the foreign agent FA
played_by FA
def=
local State:nat, Nm,IDmu,M1,Ki,Si,IDha,IDfa,SKi,PIDmu,PWmu,Kfa,SKfa,Qf,Nf,
        AFA,Nf2,BFA,Qf1,M,Nh:text
const sp1,sp2,sp3,sp4,mu_fa_nm,mu_fa_m1,fa_ha_nf,fa_mu_nf2,
        ha_fa_nh:protocol_id
init State:=2
transition
%authentication and establishment of session key phase
%Receive message {DMU,Qm,idHA,pidMU,FMU,FMS} from MU via open channel
1. State= 2∧RCV(xor((Nm'. H(IDmu.M1')),(H(H(H(IDmu.M).H(Ki)).Si))).
        H(Nm'.H(SKi. H(IDmu.M))).IDha.H(IDmu.M).xor(H(SKi.
        H(IDmu.M)),H(H(H(IDmu.M).H(Ki)).Nm')).xor(xor(H(SKi.
        H(IDmu.M)),H(H(H(IDmu.M).H(Ki)).Nm')),Si))=|>
State':=6∧secret({PIDmu},sp1,{MU,HA})∧secret({IDmu,PWmu,M},sp2,{MU})
        ∧ secret({Ki,SKi},sp3,{HA})∧secret({Kfa,SKfa},sp4,{FA,HA})
        ∧ Nf:=new()∧Qf:= H(H(Nm.H(SKi. H(IDmu.M))).Nf.SKfa)
        ∧AFA':=xor(H(SKfa.Kfa),Nf)
        % send message {MSG,Qf,AFA,idFA} to HA via open channel
        ∧SND(xor((Nm. H(IDmu.M1)),(H(H(H(IDmu.M).H(Ki)).H(SKi.
        H(IDmu.M))))).H(H(IDmu.M1).Nm.H(SKi. H(IDmu.M))).
        IDha.H(IDmu.M).xor(H(SKi. H(IDmu.M)),H(H(H(IDmu.M).H(Ki)).
        Nm)).xor(xor(H(SKi. H(IDmu.M)),H(H(H(IDmu.M).H(Ki)).Nm)),Si).
        Qf.AFA'.IDfa)∧witness(FA,HA,fa_ha_nf,Nf)
% Receive message {AHA,Nh} from HA via open channel
4. State=6∧RCV(xor((xor(H(H(IDmu.M1).H(Ki)), H(H(IDmu.M).H(Ki))).
        xor(H(SKi. H(IDmu.M1')), H(H(IDmu.M).H(Ki))).H(SKi.
        H(IDmu.M)).Nm'),H(SKfa'.Kfa'.Nh')).Nh')=|>
State':=8∧ Nf2':=new()∧BFA':= xor((xor(H(H(IDmu.M1).H(Ki)), H(H(IDmu.M).H(Ki))).
        xor(H(SKi. H(IDmu.M1)), H(H(IDmu.M).H(Ki))).Nf2'),H(SKi.
        H(IDmu.M)).Nm))∧Qf1':=H(xor((Nm. H(IDmu.M1)),(H(H(H(IDmu.M).
        H(Ki)).Si)). xor(H(H(IDmu.M1).H(Ki)), H(H(IDmu.M).H(Ki))).Nf2')
% send message {BFA,Qf1,Nf2} to MU via open channel
        ∧SND(BFA'.Qf1'.Nf2')
% FA has freshly generated random Nf2 for MU
        ∧witness(FA,MU,fa_mu_nf2,Nf2')∧request(HA,FA,ha_fa_nh,Nh')
% Receive message {Qm1} from MU via open channel
3. State=8∧RCV(H(Nm.Nf2))=|>
% FA's acceptance of the values Nm and Sn generated for FA by MU
State':=10∧request(MU,FA,mu_fa_nm,Nm)∧request(MU,FA,mu_fa_m1,M1)
end role
```

Fig. 3 Role specification for foreign agent

```
%%% role for the session %%%
role session(MU,HA,FA: agent,
          SKmuha: symmetric_key,
          H:hash_func)def= local SN1,SN2,SN3,RV1,RV2,RV3:channel(dy)
 composition
          mobileuser(MU,HA,FA,SKmuha,H,SN1,RV1)
          /\homeagent(MU,HA,FA,SKmuha,H,SN2,RV2)
          /\foreignagent(MU,HA,FA,H,SN3,RV3)
end role
%%% Role for the goal and environment %%%
 role environment()
def=
          const mu,ha,fa: agent,
          skmuha: symmetric_key,
          h : hash_func,
          idha,idfa: text,
          mu_fa_w,fa_ha_x,fa_mu_y,ha_fa_z: protocol_id,
          sp1,sp2,sp3,sp4: protocol_id
          intruder_knowledge= {mu,ha,fa,h,idfa,idha}
          composition
          session(mu,ha,fa,skmuha,h)
          /\session(i,ha,fa,skmuha,h)
          /\session(mu,i,fa,skmuha,h)
          /\session(mu,ha,i,skmuha,h)
end role
goal    secrecy_of sp1, sp2, sp3, sp4
   authentication_on fa_ha_nf,fa_mu_nf2
   authentication_on mu_fa_nm,mu_fa_m1
   authentication_on ha_fa_nh
end goal
environment()
```

Fig. 4 Role specification for session and environment

the safety of our scheme, as depicted in Fig. 5, Fig. 6 respectively. We used the security protocol animator SPAN [27] to simulate our protocol for AVISPA. The OFMC and CLAtSe back ends can be executed for replay attack checking to verify whether a legal entity can execute the protocol by searching for a passive intruder. To do this, the passive intruder (denoted by notation i) is allowed in the role environment to execute some sessions with other legal entities. Furthermore, the OFMC and CLAtSe back ends verify whether our suggested scheme is sound against the man-in-the-middle attack by the intruder (i) in the DY model checking. To do this, the intruder is allowed to know all useful public parameters of our protocol and then allowed to execute some sessions with other legal entities. Figures 5 and 6 show the simulation results for CLAtSe and OFMC back end, respectively. The result shows that our scheme is safe against main in middle attack and replay attack under CLAtSe and OFMC back end. Other two back-ends SAT-based Model-Checker (SATMC) and Tree Automata-based Protocol Analyzer (TA4SP) are not used for execution in our scheme as exclusive OR (XOR) operation is not executable under these.

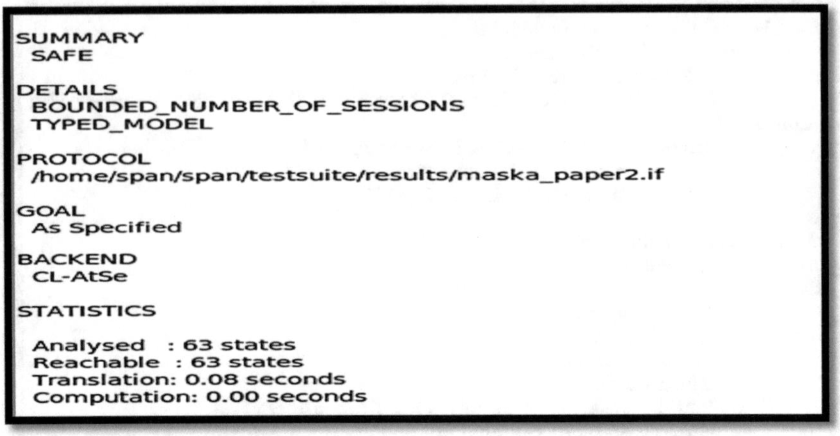

```
SUMMARY
  SAFE

DETAILS
  BOUNDED_NUMBER_OF_SESSIONS
  TYPED_MODEL

PROTOCOL
  /home/span/span/testsuite/results/maska_paper2.if

GOAL
  As Specified

BACKEND
  CL-AtSe

STATISTICS

  Analysed   : 63 states
  Reachable  : 63 states
  Translation: 0.08 seconds
  Computation: 0.00 seconds
```

Fig. 5 Result of AVISPA simulation using CL-Atse as backend

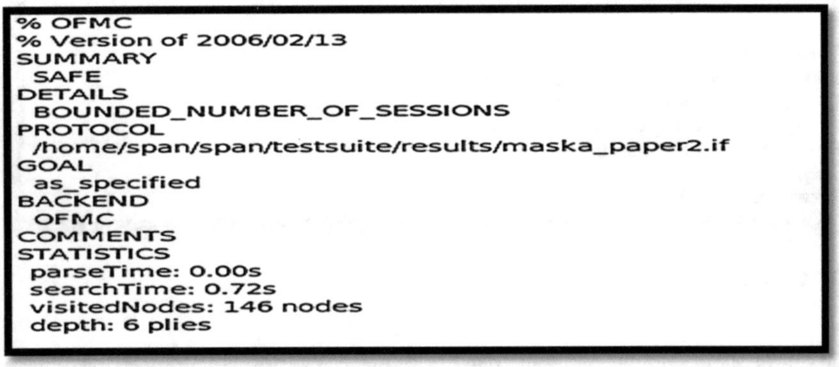

```
% OFMC
% Version of 2006/02/13
SUMMARY
  SAFE
DETAILS
  BOUNDED_NUMBER_OF_SESSIONS
PROTOCOL
  /home/span/span/testsuite/results/maska_paper2.if
GOAL
  as_specified
BACKEND
  OFMC
COMMENTS
STATISTICS
  parseTime: 0.00s
  searchTime: 0.72s
  visitedNodes: 146 nodes
  depth: 6 plies
```

Fig. 6 Result of AVISPA simulation using OFMC as backend

5 Performance Analysis and Comparison

In this segment, we compare the goodness of the proposed protocol with respect to other existing related protocols, such as the schemes of Mun et al. [5], Gope et al. [17] and Lee et al. [21].

Table 2 Security parameters comparison of our protocol with [5, 17, 21]

Security parameters	Mun et al. [5]	Gope et al. [17]	Lee et al. [21]	Proposed
SP1	No	No	Yes	Yes
SP2	No	Yes	No	Yes
SP3	No	Yes	Yes	Yes
SP4	No	No	Yes	Yes
SP5	No	Yes	Yes	Yes
SP6	Yes	No	Yes	Yes
SP7	No	No	No	Yes
SP8	No	No	No	Yes
SP9	No	Yes	No	Yes
SP10	No	No	No	Yes

5.1 Security Parameters Comparison

We have examined the protocols of Mun [5], Gope [17], Lee [21] for several security parameters (SP). We realized that all these protocols have certain demerits under different network security attacks like offline password guessing attack, impersonation attack, replay attack, etc. and even do not produce perfect mutual authentication, user anonymity, untraceability, etc. Our proposed protocol is resisting all this attack as depicted in Table 2.

SP1: user anonymity and untraceability, SP2: perfect mutual authentication, SP3: man in middle attack, SP4: replay attack, SP5: insider attack, SP6: stolen smart-card attack, SP7: impersonation attack, SP8: offline password guessing attack, SP9: perfect forward secrecy, SP10: local password verification.

Yes: Maintains security parameters; No: Does not maintain security parameters.

5.2 Computational Overhead Comparison

In this section, we analyze and compare computation time of our protocol with [5, 17, 21]. The different cryptographic functions used in [5, 17, 21] and our protocols are one-way hash function, symmetric key encryption–decryption, message authentication code, and ECC curve multiplication. The execution time for these functions is represented by the notations $T_H, T_{SYM}, T_{MAC}, T_{ECC}$, respectively. We used the experimental results recorded in [28] to obtain the total computation time for the cryptographic functions required in this section. The estimated computation times reported in [28] are given in Table 3.

Table 4 depicted the comparison of computational overhead between our scheme and Mun [5], Gope [17] and Lee [21] protocol. From Table 4, we can conclude that our scheme takes comparatively less time than Mun et al., and Gope et al. but slightly

Table 3 The estimated computation time of cryptographic functions

Description	Notation	Rough execution time (in millisecond)
One-way hash function	T_H	0.5
Symmetric encryption–decryption function	T_{SYM}	8.7
ECC point multiplication	T_{ECC}	63.075
Message authentication code	T_{MAC}	0.5

Table 4 Computational overhead comparison of our protocol with [5, 17, 21]

	Mun et al. [5]	Gope et al. [17]	Lee et al. [21]	Proposed
MU	$4T_H + 2T_{ECC} + 1T_{MAC}$	$4T_H + 1T_{SYM}$	$10T_H$	$12T_H$
FA	$3T_H + 2T_{ECC} + 1T_{MAC}$	$2T_H + 2T_{SYM}$	$8T_H$	$8T_H$
HA	$3T_H$	$7T_H + 3T_{SYM}$	$10T_H$	$12T_H$
Total	$10T_H + 4T_{ECC} + 2T_{MAC}$	$13T_H + 6T_{SYM}$	$28T_H$	$32T_H$
Rough estimation (in millisecond)	258.3	58.7	14	16

more time than Lee et al. If we consider the security parameters SP2, SP7 to SP10 as shown in Table 2, then we realized that although Lee's scheme takes less time than our scheme, it has weakness under different security attacks whereas our scheme successfully defend all these attacks.

6 Conclusions

In this paper, we perform literature survey of several research papers that were written based on mobile user authentication in GLOMONET. But we realized none of these papers have withstood under numerous network security attacks. Then we designed our own protocol. Our protocol not only withstands all kinds of known attacks, but also takes less time compare to other protocols of recent time. Furthermore, we use one-way hash function and exclusive OR operation for designing the scheme. One-way hash function takes very small time to execute and it is also far less than the other cryptographic operations. Therefore, our scheme can be useful for low power mobile device in GLOMONET.

References

1. Wu, C.C., Lee, W.B., Tsaur, W.J.: A secure authentication scheme with anonymity for wireless communications. IEEE Commun. Lett. **12**(10), 722–723 (2008)
2. Zeng, P., Cao, Z., Choo, K.-K. R., Wang, S.: On the anonymity of some authentication schemes for wireless communication. IEEE Commun. Lett. **13**(3), March 2009
3. Chang, C.-C., Lee, C.-Y., Chiu, Y.-C.: Enhanced authentication scheme with anonymity for roaming service in Global Mobility network. Comput. Commun. **32**(4), 611–618 (2009)
4. Zhou, T., Xu, J.: Provable secure authentication protocol with anonymity for roaming service in global mobility networks. Comput. Netw. **55**(1), 205–213 (2011)
5. Mun, H., Han, K., Lee, Y.S., Yeun, C.Y., Choi, H.H.: Enhanced secure anonymous authentication scheme for roaming service in global mobility networks. Math. Comput. Model. **55**(1–2), 214–222 (2012)
6. Hsieh, W.B., Leu, J.S.: Anonymous authentication protocol based on elliptic curve Diffie–Hellman for wireless access networks. Wireless Commun. Mob. Comput. **14**, 995–1006 (2014). June 2012 in Wiley https://doi.org/10.1002/wcm.2252
7. Kim, J., Kwak, J.: Improved secure anonymous authentication scheme for roaming service in global mobile network. Int. J. Security Appl. **6**(3), 45–54 (2012)
8. Jiang, Q., Ma, J., Li, G., Yang, L.: An enhanced authentication scheme with privacy preservation for roaming services in global mobility networks. Wireless Pers. Commun. **68**(4), 1477–1491 (2013)
9. Lee, T.-F.: User authentication scheme with anonymity, unlinkability and untrackability for global mobility networks. Commun. Netw. **6**(11), 1404–1413 (2013)
10. He, D., Zhang, Y., Chen, J.: Cryptanalysis and improvement of an anonymous authentication protocol for wireless access networks. Wireless Pers. Commun. **74**(2), 229–243 (2014)
11. Wen, F., Susilo, W., Yang, G.: A secure and effective user authentication scheme for roaming service in global mobility networks. Wireless Pers. Commun. **73**(3), 993–1004 (2013)
12. Zhao, D., Peng, H., Li, L., Yang, Y.: A secure and effective anonymous authentication scheme for roaming service in global mobility networks. Wireless Pers. Commun. **78**, 247 (2014). https://doi.org/10.1007/s11277-014-1750-y
13. Kuo, W.-C., Wei, H.-J., Cheng, J.-C.: An efficient and secure anonymous mobility network authentication scheme. J. Inf. Secur. Appl. **19**, 18–24 (2014)
14. Gope, P., Hwang, T.: Enhanced secure mutual authentication and key agreement scheme preserving user anonymity in global mobile networks. Wireless Pers. Commun. **82**, 2231 (2015). https://doi.org/10.1007/s11277-015-2344-z
15. Gope, P., Hwang, T.: Lightweight and energy-efficient mutual authentication and key agreement scheme with user anonymity for secure communication in global mobility networks. IEEE Syst. J. **10**(4), December 2016
16. Karuppiah, M., Kumari, S., Das, A.K., Li, X., Wu, F., Basu, S.: A secure lightweight authentication scheme with user anonymity for roaming service in ubiquitous networks. Secur. Commun. Netw. **9**(17), 4192–4209 (2016)
17. Gope, P., Hwang, T.: An efficient mutual authentication and key agreement scheme preserving strong anonymity of the mobile user in global mobility networks. J. Netw. Comput. Appl. **62**, 1–8 (2016)
18. Xu, L., Wu, F.: A novel three-factor authentication and key agreement scheme providing anonymity in global mobility networks. Secur. Commun. Netw. **9**(16), 3428–3443 (2016)
19. Arshad, H., Rasoolzadegan, A.: A secure authentication and key agreement scheme for roaming service with user anonymity. Int. J. Commun. Syst. **30**(18), e3361 (2017)
20. Madhusudhan, R., Suvidha, K.S.: An efficient and secure user authentication scheme with anonymity in global mobility networks. In: Proceedings of the 31st International Conference on Advanced Information Networking and Applications Workshops (WAINA), pp. 19–24, Mar 2017

21. Lee, C.-C., Lai, Y.-M., Chen, C.-T., Chen, S.-D.: Advanced secure anonymous authentication scheme for roaming service in global mobility network. Wireless Pers. Commun. **94**(3), 1281–1296 (2017)
22. Park, K., Park, Y., Park, Y., Reddy, A.G., Das, A.K.: Provably secure and efficient authentication protocol for roaming service in global mobility networks. IEEE Access. https://doi.org/10.1109/access.2017.2773535
23. Kocher, P., Jaffe, J., Jun, B.: Differential power analysis. In: Proceedings of the 19th Annual International Cryptology Conference, Santa Barbara, CA, USA, pp. 388–397, August 1999
24. Messerges, T.S., Dabbish, E.A., Sloan, R.H.: Examining smartcard security under the threat of power analysis attacks. IEEE Trans. Comput. **51**(5), 541–552, May 2002
25. AVISPA. Automated Validation of Internet Security Protocols and Applications. Available: http://www.avispa-project.org/
26. von Oheimb, D.: The high-level protocol specification language HLPSL developed in the EU project AVISPA. In: Proceedings of the 3rd APPSEM, Frauenchiemsee, Germany, pp. 1–17 (2005)
27. AVISPA. SPAN, The Security Protocol Animator for AVISPA. Available: http://www.avispa-project.org
28. Banerjee, S., Odelu, V., Das, A.K., Chattopadhyay, S., Kumar, N., Park, Y., Tanwar, S.: Design of an anonymity-preserving group formation based authentication protocol in global mobility networks. IEEE Access **6**, 20673–20693 (2018). https://doi.org/10.1109/ACCESS.2018.2827027

Efficient Entity Authentication Using Modified Guillou–Quisquater Zero-Knowledge Protocol

Debasmita Dey and G. P. Biswas

Abstract Fiat and Shamir were the first ones to come up with the idea of proving a zero-knowledge scheme with the concept of knowledge about knowledge. However, their scheme required to be run for a considerable amount of time, so that the probability of a false claimant being correct is less. This meant greater computation cost. Taking inspirations and motivations from Fiat and Shamir, Guillou and Quisquater developed a scheme based on the same concept of knowledge about knowledge. Their scheme was based on the working of RSA, and it required to be run fewer number of times for attaining the same probability of a false claimant being correct. However, some shortcomings have been observed in Guillou–Quisquater's scheme. In this work, we pointed out those shortcomings and tried to modify their scheme. The modified scheme is developed on the difficulty of Diffie–Hellman key exchange protocol. The proposed scheme also requires to be run for lesser number of rounds. We also performed the security analysis and performance analysis for the proposed scheme.

Keywords Zero-knowledge protocol · Fiat–Shamir protocol · Guillou–Quisquater protocol · Diffie–Hellman key exchange

1 Introduction

Zero-knowledge proof holds an important position in public-key cryptography. It is generally considered to be a refined technique to restrict the extent of data sent from one party to another in a cryptographic scheme. This protocol is used in many situations like authorization for accessing a server, exchange of communication between clients and banks, digital signatures, and a lot more. A recent article showed that zero-knowledge protocol has been very useful in providing secure communication,

D. Dey (✉) · G. P. Biswas
Indian Institute of Technology (Indian School of Mines), Dhanbad, India
e-mail: debasmita.dey25@gmail.com

G. P. Biswas
e-mail: gpbiswas@gmail.com

© Springer Nature Singapore Pte Ltd. 2020
M. Chakraborty et al. (eds.), *Proceedings of International Ethical Hacking Conference 2019*, Advances in Intelligent Systems and Computing 1065, https://doi.org/10.1007/978-981-15-0361-0_15

privacy, and authentication. In recent days, where cryptocurrencies are gaining much popularity, Zerocoin, Monero, Zcash, and PIVX are some of the platforms related to cryptocurrencies that call for the involvement of this scheme. Identity verification like authenticating a PIN by a security component can hold private information without factorizing or revealing them. Remote-control systems and other microprocessor-based devices generally use this scheme.

The definition of zero-knowledge protocol is often stated as, there lie two groups the claimant and the verifier, where the claimant claims that he knows a value but without actually revealing the value to the verifier. We will discuss the details of these two entities (claimant and verifier) in later sections. Zero-knowledge-based smart cards developed by Burmester et al. help in securing identification to avoid fraudulent actions [1]. In 2016, a unique approach of inspecting whether an object is a nuclear weapon without any kind of internal investigation which may reveal secret information about the object; this innovative discovery was performed by the Princeton Plasma Physics Laboratory and Princeton University.

Goldwasser et al. [2] first introduced the concept of zero-knowledge proof of identity in 1985, their work depicted proofs based on the query whether any input x is a member of a language L, the idea that was mainly used by the authors was of revealing a single bit of information by the claimant. An ample amount of research related to this scheme and interactive proof of identity has been done by Goldwasser and Sipser [3]; Brassard and Crepeau [4] worked on the non-transitive transfer of confidence to name a few. Later, in 1986 Feige, Fiat, and Shamir [2] introduced a practical scheme based on the difficulty of factoring and computing the square root modulo a large composite number, here they adopted a new concept, where the claimant does not reveal any information whatsoever but proves to the verifier that he has knowledge about the information. Guillou and Quisquater [5] published an RSA-based zero-knowledge protocol in 1988, where they proved of requiring less storage that is performing the operations lesser number of times than that of Feige et al. [2] however, they mentioned that each operation requires a greater number of calculations. Later on, in 1989, Schnorr [6] found out a way of proving the knowledge about discrete logarithm using zero-knowledge protocol. Over the few decades, zero-knowledge protocol has been used in a considerable amount of work [7–13] in various applications. For years, this methodology has been used in various cryptographic methods like encryption schemes secured against chosen ciphertext attack by Naor and Yung [14] in 1990. Fujisaki et al. [15] used zero-knowledge protocol for modular polynomial relations in 1997. Bangerter et al. [16] performed a zero-knowledge proof for exponentiation-based discreet logarithm. Fiat and Shamir scheme's security was discussed by Kalai et al. [17], whereas Ezziri et al. [18] provided a variation of Guillou–Quisquater scheme. Peikert [19] provided for the non-interactive zero-knowledge for the NP hard problem using learning with errors in 2019.

With this paper, we bring out a variation of Guillou and Quisquater [5] protocol, where unlike running the algorithm several times for confirming the authenticity of the claimant, the same assurance can be achieved by running the algorithm lesser number of times. However, some shortcomings have been noticed in the scheme, we will point out that precisely and perform the required modification. In a world, where time and computation hold high cost executing complex computation for several times increases the overhead, hence the need for more efficient schemes.

1.1 Objective of the Paper

In the introduction, given above, we see that considerable amount of work has been done based on zero-knowledge protocol; however, very few works have been proposed for improving the zero-knowledge protocol. Guillou–Quisquater took initiative in providing efficient scheme, which requires fewer number of steps; however, there exist some flaws. Through this paper, we achieve the following goals:

- We have discussed some existing schemes and performed their cryptanalysis. We laid out the ways in which the security or soundness of these schemes may be jeopardized.
- We proposed our new scheme for zero-knowledge proof of identity, which uses the concept of knowledge about knowledge.
- The three paradigms namely completeness, soundness, and zero-knowledge for proving a scheme to be zero-knowledge secure is provided.

2 Background and Definitions

We lay out the commonly used methods in this section that is being used in our paper. The topics included here are zero-knowledge protocol, claimant, and verifier.

2.1 Zero-Knowledge Protocol

According to Fiege, Fiat, and Shamir, the concept of knowledge about knowledge is more important than mere knowledge. So, the new definition suggests the condition where a prover or claimant claims that he has knowledge about an identity for which he first sends a witness. The verifier in turn sends a challenge, which is needed to be answered by the claimant. The verifier then verifies whether the response is valid or not. Mathematically, $f(c, r) = 1$, where c is the challenge and r is the response, f is the function that computes the validity of c and r, if f is 1, the claimant is accepted otherwise not.

2.2 Claimant

The party that aims to prove its knowledge as well as identity is referred to as the claimant (C hereinafter). C varies from being a person, server, client, or process.

2.3 Verifier

The party that verifies the claims of C in any communication is referred to as the Verifier (V hereinafter). V can create some proof by himself for C.

3 Discussion of Previous Scheme

In this section, we would like to bring out the discussion of previously existing schemes about zero-knowledge protocol. We brief out Fiat–Shamir [2] protocol and Guillou and Quisquater [5] protocol, describing each computation elaborately.

3.1 Fiat–Shamir Scheme

As mentioned earlier, it was Fiat and Shamir [2] first who came up with the concept of knowledge about knowledge. They considered the challenge value between 0 and 1, thus making the probability of a false claimant being correct $1/2$. This value decreases with each subsequent round. Therefore, this scheme calls for greater number of rounds for reliable results.

3.1.1 Primary Considerations

Trusted authority (*TA* hereinafter) chooses two large prime numbers p and q and calculates the modulus value $N = p \times q$. Here, *TA* announces N to be public and p and q are kept secret.

C selects a secret integer s, such that $s \in_R \{1, \ldots, N-2\}$ and computes a value

$$v \equiv s^2 \bmod N \tag{3.1.1}$$

keeping r as secret key and v as public key is shared with *TA*.

3.1.2 Verification

Verification is done by V in four exchanges, which includes witness, challenge, response, and verification.

C chooses a random number referred to as the commitment r, whose range varies from 0 to $N - 1$. C then calculates the witness

$$w = r^2 \bmod N \qquad (3.1.2)$$

and sends w to V.

V in turn sends C a challenge ch such that ch is either 0 or 1. C responds to V by multiplication of the random r and the exponentiation of the secret key s with the challenge ch. Therefore, we get a response

$$R \equiv r \times s^{ch} \bmod N \qquad (3.1.3)$$

The verifier checks whether the two values of R^2 are congruent to wv^{ch} with respect to the modulus of N. Computations of congruency checking are given below,

From Eq. 3.1.3, $R^2 = (r \times s^{ch})^2$.

Expanding we get, $R^2 = r^2 \times s^{2ch} \bmod N = r^2 \times (s^2)^{ch} \bmod N$.

Substituting values from Eqs. 3.1.1 and 3.1.2, we get,

$R^2 = wv^{ch} \bmod N$; hence, the verifier is convinced that claimant knows the secret key for this particular round. This verification is performed several times with the challenge value being 1 or 0.

3.2 Guillou–Quisquater Scheme

Fiat–Shamir's [2] work has been extended by Guillou and Quisquater [5]. Unlike [2] they attempted to increase the range of the challenge which decreases the probability of C being correct. As mentioned earlier, they adopted the methodologies of RSA public-key cryptosystem. In this protocol, C attempts to prove his knowledge about the private value z, where z holds a relation with the public value v and the public key e as

$$z^e \times v \equiv 1 \bmod N \qquad (3.2.1)$$

here, N is the modulus value of RSA.

3.2.1 Preliminary Considerations

Primarily, two prime integers of larger number of bits p and q are chosen. It is taken care of that $p \neq q$. N is calculated as the product of the two prime values p and q as

$N = p \times q$. p and q are kept secret, whereas N is made public by a trusted authority (*TA* hereinafter).

The Euler totient function is calculated as $\varphi(N) = (p-1) \times (q-1)$; this value is also kept secret by the *TA*.

TA decides on two integer values z and v for C and V respectively, and the relation is mentioned in Eq. (3.2.1). It must be noted here that v is the multiplicative inverse of z^e with respect to N.

The procedure consists of several rounds, where each round includes three exchanges between C and V. C must be accurate in every round for him to be authenticated otherwise the whole process is terminated.

3.2.2   Round

For the first exchange, C chooses a random number r and raises it to the power of e yielding the witness value

$$w = r^e \bmod N \text{ to } V \tag{3.2.2}$$

V in turn sends a challenge ch (an integer value) randomly, such that ch ranges from 1 to e.

C then prepares a response R by exponentiating the private value z with the challenge ch and multiplying the result with the random value r, which gives

$$R = r \times z^{ch} \tag{3.2.3}$$

The verification procedure is performed by the verifier V. Here, V tries to calculate w from the information available to him that is R, e, ch, and v. He first starts with multiplication of the exponentiation of the response R with the public value e and the exponentiation of the public value v with the challenge ch. Giving out the detailed calculation, we have,

$$R^e \times v^{ch} \bmod N \tag{3.2.4}$$

Putting the value of R from Eq. (3.2.3) in Eq. (3.2.4), we have

$$\left(r \times z^{ch}\right)^e \times v^{ch} \bmod N \tag{3.2.5}$$

which leads to

$$r^e \times z^{che} \times v^{ch} \bmod N \tag{3.2.6}$$

Now, from Eq. (3.2.1), we get $v \equiv z^{-e} \bmod N$, putting this in Eq. (3.2.6), we have $r^e \times z^{che} \times z^{-che}$, finally giving $r^e \bmod N = w$.

4 Security Analysis of the Above Schemes

In this section of the paper, we need to discuss that under what circumstances may Fiat–Shamir and Guillou–Quiaquater schemes fail and not provide acceptable results. We also provide the shortcomings and lack of clarity of Guillou–Quisquater scheme security analysis of Fiat–Shamir Scheme.

4.1 Fiat–Shamir Protocol

The challenge value here can be either 0 or 1. So, there may arise few cases, where C is deceitful but still can qualify each round like,

- If C guesses $ch = 1$, the value of witness, w is calculated as $w = r^2 v^{-1} \bmod N$ and sends w as the witness.
- Suppose the guess of $ch = 1$ is correct, C sends the response $R = r$ to the verifier. Where it can be noticed that the verifier accepts the claimant as $R^2 = wv^{ch}$.
- If C guesses $ch = 0$, he calculates the value of witness $w = r^2 \bmod N$ and sends this witness value w to the verifier.
- Suppose the guess of $ch = 0$ is correct, then C sends the response $R = r$ to the verifier and the verifier in turn accepts as $R^2 = wv^{ch}$.

It is seen here that each time, the probability of acceptance of the claims of the claimant is $1/2$; when the guess becomes incorrect, then the whole process is discarded. Therefore, due to this reason, this scheme needs to be run for several rounds for decreasing the probability of a false claimant being incorrect.

4.2 Guillou–Quisquater Protocol

In Fiat–Shamir scheme, we have observed that the claimant selects a private key and a public key, where he keeps the private key with him and gives the public key to the verifier and the TA. However, in Guillou–Quisquater scheme, this public and private value is computed by the TA, but how the claimant is getting to know the secret value is not mentioned. Secret values cannot be openly provided to any party, it must be sent as a function, so, that calculation of a function is not performed anywhere in [5].

5 Proposed Modified Scheme

With this section, we define out our proposed scheme, which like [2] needs to be run fewer number of times. We utilized Diffie–Hellman key exchange protocol and the use of large prime numbers will be evident. Our scheme also requires the involvement of any trusted authority.

5.1 Primary Consideration

A large prime integer P, whose size is greater than 512 bits is taken by the trusted authority; *TA* also takes a generator of the prime number P as g. *TA* announces both the values public.

The claimant chooses a secret value x from the range of $\{1, \ldots, P - 1\}$ and a public value is calculated by him as an exponentiation of the generator g with the chosen secret value x, with modulus P. The secret value is not published and kept with the claimant, whereas the public value is published to the verifier and the *TA*. Therefore, summing up, we have

$$\text{Private: } x \in_R \{1, \ldots, P - 1\}$$
$$\text{Public: } e = g^x \bmod P \qquad (5.1.1)$$

5.2 Verification

As a first step, the claimant chooses a random number r within the range of 1 to $P - 1$, mathematically, $r \in_R \{1, \ldots, P - 1\}$.

Next, C computes the witness by modular exponentiation of the generator g and the random integer r with respect to P. Denoting this mathematically, we have,

$$w = g^r \bmod P \qquad (5.1.2)$$

Now, the verifier V sends a challenge, whose value lies in the range of $\{1, \ldots, P - 1\}$, such that

$$ch \in_R \{1, \ldots, P - 2\}.$$

C now provides a response to V, where C first performs a modular multiplication of the challenge ch and the private key x with respect to $P - 1$. Then modular addition is performed with the previously obtained value, that is $ch \times x \bmod P - 1$ and the random value, r chosen by C with respect to $P - 1$. Therefore, the response is given as,

$$R = r + (ch \times x) \bmod P - 1 \qquad (5.1.3)$$

The verifier V, now, checks for whether the response provided is correct or not, and the claimant is true and has knowledge about the secret key x.

V starts by the modular exponentiation of the generator and the response with respect to the prime P; this would result into a product of witness and a value, v, which is the modular exponentiation of the public key and the challenge with respect to P. Thus, mathematically denoting,

Computes $g^R \bmod P$

$= g^{r+chx \bmod P-1} \bmod P$, according to Eq. (5.1.3)

$= g^{r \bmod P-1} g^{chx \bmod P-1} \bmod P$

$= w \times e^{ch} \bmod P$, from Eq. (5.1.2) and Eq. (5.1.1)

Now, here all the values are known by the verifier.

6 Security Analysis of the Proposed Scheme

In this section, we discuss the security analysis of our proposed scheme. We will provide the zero-knowledge proof, based on the three paradigms, namely completeness, soundness, and zero knowledge.

6.1 Completeness

Completeness refers to the condition, where the unbiased verifier will be convinced by an honest claimant that the statement or proposal made by the claimant is true.

In our case, we see that the verifier retrieves $w \times e^{ch} \bmod P$, all the values known to him from the response R, given to him by the claimant. The public key obtained in the above case could only have been possible if the claimant knew the private key x.

Obtaining the value of the private key from the public key involves solving the discreet logarithm problem, which is not possible. Moreover, the witness that has been given by the claimant to the verifier included the usage of some random value, known only to the claimant. So, the ultimate result would not have included witness if the claimant did not have knowledge about the correct random value r.

6.2 Soundness

For soundness, we need to prove that any dishonest claimant cannot prove his identity to the verifier except with negligible probability.

The claimant might try guessing the value of the challenge, and as mentioned above, the range of the challenge is generally taken from 1 to $P - 2$, where P is of large size (>512 bits). Therefore, the probability of guessing the bit correctly is approximately around $1/2^{512}$. Thus, it is evident that the probability is excessively negligible.

The private key x that is being used is generated only by the honest claimant. So, guessing that value also involves a negligible probability. Same goes for guessing the random value r chosen for generating the witness.

6.3   Zero Knowledge

In this proof system, we need to show that the verifier is only provided with the knowledge about the knowledge. The actual knowledge that the claimant holds is not given to the verifier.

We see in our mentioned scheme, that though the claimant is providing the witness w as well as response R to the verifier, the actual information or knowledge is not given out to the verifier. Neither the private key x nor the random number r can be extracted from the public values, due to discreet logarithm problem.

Finally, it is clear to the verifier that the claimant trying to prove his identity is honest and he has knowledge about the confidential information.

7   Performance Analysis

Measuring the performance of our mentioned scheme with the Fiat–Shamir and Guillou–Quisquater scheme, we can derive the result of the following Table 1.

From the above table, it is clear that the probability of each round of Fiat–Shamir scheme is $1/2$ as the verifier has only two options for choosing the challenge, either 0 or 1. So, to achieve a probability of 3×10^{-8}, there must be 25 rounds.

For the Guillou–Quisquater scheme, the value of e is taken as 3 most of the times; however, those schemes include padding. Therefore, considering the highest and a safe value of e, we get a probability of $1/65537$, which is an approximate value. The number of rounds required to be run for making the scheme secure is around 1; however, the probability can be decreased and hence the performance can be increased, if we increase the number of rounds.

Coming to our proposed scheme, we would like to mention here, that we have taken a value of the challenge, whose range is from 1 to 2^{512}, thus resulting in an

Table 1 Performance analysis of two schemes and the proposed modified scheme

Schemes	Probability of one round	Number of rounds required
Fiat–Shamir	$1/2$	25
Guillou–Quisquater	$\approx 1/65537$	≈ 1
Proposed scheme	$1/2^{512}$	1

approximate probability of $1/2^{512}$ and this small value of probability in a single round calls for fewer number of rounds for obtaining security.

8 Conclusion

In concluding, we would like to point out our contribution to developing a Guillou–Quisquater modified protocol related to the zero-knowledge proof of identity. We maintained the concept of Fiat–Shamir on developing schemes that give knowledge about knowledge and we also adopted the idea of Guillou and Quisquater of performing the round fewer number of times. Unlike Guillou and Quisquater, we concentrated on applying discreet logarithm problem and Diffie–Hellman key exchange protocol instead of RSA. Few works have mentioned that increasing the number of rounds in Guillou–Quisquater may decrease the probability and also using $2e$ instead of e, however, it has not been brought to anyone's attention that the details of exchange of the secret information from the trusted authority to the claimant have not been mentioned. We also performed a security analysis of our proposed modified scheme and a performance analysis of the three schemes has been provided with the probabilities of each round.

References

1. Burmester, M., Desmedt, Y., Beth, T.: Efficient zero-knowledge identification schemes for smart cards. Comput. J. **35**(1), 21–29 (1992)
2. Goldwasser, S., Micali, S., Rackoff, C.: The knowledge complexity of interactive proof systems. SIAM J. Comput. **18**(1), 186–208 (1989)
3. Goldwasser, S., Sipser, M.: Private coins versus public coins in interactive proof systems. In: Proceedings of the Eighteenth Annual ACM Symposium on Theory of Computing. ACM (1986)
4. Brassard, G., Crepeau. C.: Non-transitive transfer of confidence: a perfect zero-knowledge interactive protocol for SAT and beyond. In: 27th Annual Symposium on Foundations of Computer Science (sfcs 1986). IEEE (1986)
5. Guillou, L.C., Quisquater, J.-J.: A Practical Zero-Knowledge Protocol Fitted to Security Microprocessor Minimizing both Transmission and Memory. Workshop on the Theory and Application of Cryptographic Techniques. Springer, Heidelberg (1988)
6. Schnorr, C.-P.: Efficient identification and signatures for smart cards. In: Conference on the Theory and Application of Cryptology. Springer, New York (1989)
7. Goldreich, O., Micali, S., Wigderson, A.: Proofs that yield nothing but their validity or all languages in NP have zero-knowledge proof systems. J. ACM (JACM) **38**(3), 690–728 (1991)
8. Rackoff, C., Simon, D.R.: Non-interactive zero-knowledge proof of knowledge and chosen ciphertext attack. In: Annual International Cryptology Conference. Springer, Heidelberg (1991)
9. Cramer, R., Damgård, I., MacKenzie, P.: Efficient Zero-Knowledge Proofs of Knowledge Without Intractability Assumptions. International Workshop on Public Key Cryptography. Springer, Heidelberg (2000)

10. Cramer, R., Damgård, I.: Zero-knowledge proofs for finite field arithmetic, or: can zero-knowledge be for free? In: Annual International Cryptology Conference. Springer, Heidelberg (1998)
11. Tompa, M., Woll, H.: Random self-reducibility and zero knowledge interactive proofs of possession of information. In: 28th Annual Symposium on Foundations of Computer Science (sfcs 1987), IEEE (1987)
12. Bellare, M., Goldwasser, S.: New paradigms for digital signatures and message authentication based on non-interactive zero knowledge proofs. In: Conference on the theory and application of cryptology. Springer, New York (1989)
13. Kilian, J., Micali, S., Ostrovsky, R.: Minimum resource zero knowledge proofs. In: 30th Annual Symposium on Foundations of Computer Science, IEEE (1989)
14. Naor, M., Yung, M.: Public-key cryptosystems provably secure against chosen ciphertext attacks. In: Proceedings of the Twenty-Second Annual ACM Symposium on Theory of Computing. ACM (1990)
15. Fujisaki, E., Okamoto, T.: Statistical zero knowledge protocols to prove modular polynomial relations. In: Annual International Cryptology Conference. Springer, Heidelberg (1997)
16. Bangerter, E., et al.: Automatic generation of sound zero-knowledge protocols. In: IACR Cryptology ePrint Archive 2008, 471 (2008)
17. Kalai, Y.T., Rothblum, G.N., Rothblum, R.D.: From obfuscation to the security of Fiat-Shamir for proofs. In: Annual International Cryptology Conference. Springer, Cham (2017)
18. Ezziri, S., Khadir, O.: Variant of Guillou-Quisquater zero-knowledge scheme. Int. J. Open Problems Comput. Sci. Math. 11(2) (2018)
19. Peikert, C., Shiehian, S.: Noninteractive zero knowledge for np from (plain) learning with errors. In: Cryptology ePrint Archive, Report 2019/158 (2019). https://eprint.iacr.org/2019/158

Ethical Hacking: Redefining Security in Information System

Sanchita Saha, Abhijeet Das, Ashwini Kumar, Dhiman Biswas
and Subindu Saha

Abstract On defining the severe status of information security in the present world, we come across a very renowned technical term known as 'ethical hacking'. Ethical hacking refers to the art of unmasking the vulnerabilities and the weakness in a computer or an information system. The process involves duplication of intents and actions of other malevolent hackers. Ethical hacking can be also called as 'penetration testing', 'intrusion testing', or 'red teaming'. Talking about the term 'hacking', it is basically a challenging and an invigorating procedure to steal information from an unknown computer system or may be a device without the prior knowledge of the owner of that system. Now by the term 'ethical', we understand the process of hacking is done for an ethical purpose which will result in a boon for the society. An ethical hacker tries to recover or destroy the stolen information or data by the non-ethical hackers. The process of hacking can thus become a boon as well as a curse for the society, and it depends upon the intention of a hacker. This is no doubt that a very strong procedure and severely based on what way it is used. This paper elicits the various methodologies and concepts related to ethical hacking as well as the tools and software used in the process along with the future aspects and emerging technologies at this field.

Keywords Ethical · Security · Vulnerabilities · Hacker · Intrusion

S. Saha (✉) · A. Das · A. Kumar
Haldia Institute of Technology, Haldia, India
e-mail: sanchita.cse2007@gmail.com

A. Das
e-mail: abhijeetd720@gmail.com

A. Kumar
e-mail: ashwinikumaredu@gmail.com

D. Biswas
South Calcutta Polytechnic College, Kolkata, India
e-mail: bisdhi24@gmail.com

S. Saha
Institute of Engineering and Management, Kolkata, India
e-mail: subindu.saha@iemcal.com

© Springer Nature Singapore Pte Ltd. 2020
M. Chakraborty et al. (eds.), *Proceedings of International
Ethical Hacking Conference 2019*, Advances in Intelligent
Systems and Computing 1065, https://doi.org/10.1007/978-981-15-0361-0_16

1   Introduction

Ethical hacking can be defined as an ultimate security professional work commonly termed as 'white hat hacking'. Ethical hackers are well known to detect and exploit vulnerabilities and weakness out of various systems. An ethical hacker uses those skills in a legitimate and a lawful manner to find out the vulnerabilities existing in a system and fix them before the malicious activists try to break in through.

An ethical hacker role is similar to a penetration tester, but it involves bigger duties. They break into systems legally and ethically. This is the primary difference of legality between ethical hackers and non-ethical hackers.

Apart from testing processes, ethical hackers are associated with several other responsibilities. The main idea is to imitate a malicious hacker [1] at work, and rather than exploiting the susceptibilities for malicious purposes, they seek countermeasures to shore up the systems' defence. An ethical hacker might employ all or some of these strategies to enter into a system:

- Scanning Ports and Seeking Vulnerabilities: An ethical hacker uses port scanning tools like Nmap or Nessus to scan one's system and locate the open ports. The vulnerabilities with each of the ports can be studied and remedial actions can be taken.
- An ethical hacker examines the patch installations and creates prevention to save them getting exploited.
- The ethical hacker may get involved in social engineering concepts like dumpster diving rummaging through trash bins for passwords, sticky notes, charts, or other things with vital information [2] that can generate an attack (Fig. 1).

Fig. 1   A real-time image of threat model

Fig. 2 Threat model of ethical hacking

- An ethical hacker may also employ other social engineering techniques like shoulder surfing to gain access to crucial information or play the kindness card to trick employees to part with their password.
- An ethical hacker will attempt to evade intrusion prevention systems (IPS), intrusion detection systems (IDS), honeypots, and firewalls.
- Bypassing and cracking wireless encryption, sniffing networks and capture Web servers and Web applications.
- Ethical hackers may also handle issues related to device theft and employee fraud as well as solving the problems with locating the lost devices and unlocking through bypassing the password-protected devices and help the cyber units (Fig. 2).

A real-time image has taken for just to show that each and every second our personal information is compromised and how can we detect that well, by knowing the threats.

Therefore, we must be aware of the threat modelling.

Threat modelling helps us to know about the attacks and which appropriate steps can be taken in order to protect our information.

2 Types of Hacking

We can separate out hacking into different categories, based on what is being hacked. Here, it contains a set of examples:

- Website Hacking: Hacking a website means enchanting unauthorized control over a Web server and its related software such as databases and other interfaces.

- Network Hacking: Hacking a network means gathering information about a network by using different tools like Telnet, nslookup, ping, TRACERT, and netstat, with the resolved to harm the network system and hamper its operation.
- Email Hacking: This contains getting unauthorized access on an email account to using it.
- Password Hacking: This is the process of recuperating secret passwords from data that has been stored in or diffused by a computer system.
- Computer Hacking: This is the process of stealing computer ID and password [3] by applying hacking methods and getting unauthorized access to a computer system.
- Ethical Hacking: Ethical hacking involves finding weaknesses in a computer or network system for testing purpose and finally getting them fixed on it. It is very much comparable to penetration testing.

2.1 Penetration Testing

This is basically a replicated cyber-attack against your computer system to check for the exploitable susceptibilities present in the system. In the context of Web application security, penetration testing is commonly used to enhance a Web application firewall (WAF).

Penetration testing comprises the attempted piercing of any number of application systems, (e.g. application protocol interfaces (APIs), front-end/back-end servers) to uncover vulnerabilities, such as unsensitized inputs that are susceptible to code injection attacks (Fig 3).

The pen testing process can be broken down into the following five stages:

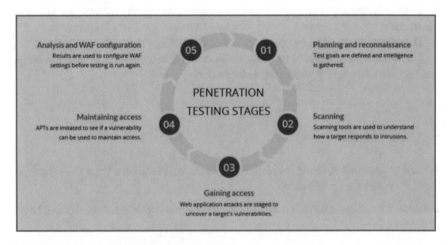

Fig. 3 Penetration testing stages of ethical hacking

2.1.1 Planning and Reconnaissance

- The scope and goals of a test, including the systems to be addressed and the testing methods to be used, are significant.
- Intelligence (e.g. network and domain names, mail server) is required to better understand how a target works and its probable vulnerabilities.

2.1.2 Scanning

The next is to understand how the goal application will respond to various invasion attempts. This is characteristically done using:

- Static analysis: Reviewing an application's code to estimate the way it performs while running. Tools can scan the whole of the code in a single pass.
- Dynamic analysis: Reviewing an application's code in a running state. This is a more real way of scanning [4], as it delivers a real-time view into an application's performance.

2.1.3 Gaining Access

This phase uses Web application attacks, such as cross-site scripting, SQL injection, and back doors, to uncover a target's vulnerabilities. Testers then try and exploit these vulnerabilities, stealing data, intercepting track, etc., to understand the damage they can cause.

2.1.4 Maintaining Access

The target of this stage is to see if the vulnerability can be used to achieve a tenacious presence in the exploited system long enough for a bad actor to gain in-depth access. This concept is to imitate unconventional persistent threats, which often remain in a system for months in order to steal an organization's most crucial data.

2.1.5 Analysis

The results of the penetration test are then assembled into a report listing:

- Precise susceptibilities that were exploited.
- Delicate data that was accessed.
- The certain amount of time the pen tester was able to remain undetected in the system.

The above information is scrutinized by security personnel to help configure an enterprise's WAF settings and other application security solutions to patch suscepti-bilities and protect against future attacks.

3 Tools Used for Ethical Hacking

3.1 Nmap

Nmap or a Network Mapper is a free open-source tool which widely used in the purpose of network detection and security checking. It was originally built to scan large networks, but it can work equally for single hosts as well. It makes it easy for the network administrators for the tasks such as network inventory, monitoring host, or service up-time and upgrade schedules managing service. Nmap [5] works well in the well-known operating systems such as the Windows, Mac OS X, as well as it is already installed in Kali Linux platform and BackTrack.

Nmap basically uses the raw IP packets to determine these various terminologies used in cybersecurity.

- The hosts which are available on the network.
- The operating systems they are running on.
- The services offered by those hosts.
- The type of firewalls in use and other such characteristics.

3.2 Burp Suite

A Burp Suite is a popular platform that is widely used for performing security testing of Web applications. It has several tools that works with the collaboration of whole testing process support, from initial mapping and analysis of an application's attack surface, to finding and manipulating security vulnerabilities. Burp Suite [6] can be easily operated on and it provides the administrators full control to combine advanced manual techniques with automation for effectual testing. Burp Suite can be simply configured, and it contains specific features to assist even the most experienced testers with their work (Fig. 4).

This is an original screenshot of the working tool Burp Suite along with the XAMPP control panel opened on the window.

Step 1: Go to proxy tab and then select Intercept → click on Intercept and make it enable.
Step 2: Open Web browser and open any website to capture its traffic and vulnera-bilities.
Step 3: The vulnerabilities are detected in the Burp Suite Panel.

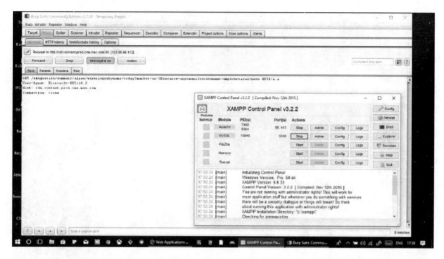

Fig. 4 Working tool of Burp Suite along with XAMPP control panel

3.3 Maltego

Maltego [7] is an interactive data mining tool which extracts directed graphs for link analysis. For online investigations, this tool verdicts the relationship between pieces of information from numerous sources which are located on Internet (Fig. 5).

Fig. 5 Vulnerability detection in Burp Suite panel

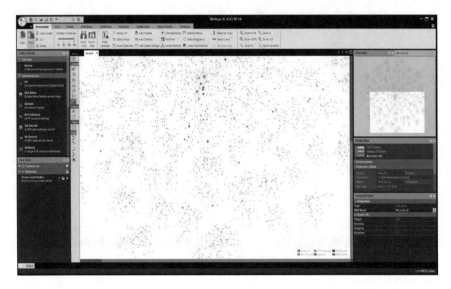

Fig. 6 A directed graph of Maltego for link analysis

- It uses the idea of transmutes to automate the process of interrogating different data sources, and this information is then displayed on a node-based graph suited for performing link analysis.
- Presently, there are three versions of the Maltego client namely Maltego CE, Maltego Classic, Maltego XL.
- All these Maltego clients come with access to a library of standard transforms for the discovery of data from an eclectic range of public sources that are usually used in online digital forensics and online investigations (Fig. 6).
- Because Maltego can effortlessly integrate with nearly any data source, many data vendors have chosen to use Maltego as a delivery platform for their private data. This also means Maltego can be adapted to someone's own, unique supplies (Fig. 7).

3.3.1 What Does Maltego Do?

The focus of using Maltego is evaluating real-world relationships between information and knowledge which is publicly accessible on the Internet. This includes footprinting Internet infrastructure as well as information gathering about the people and the corresponding organization.

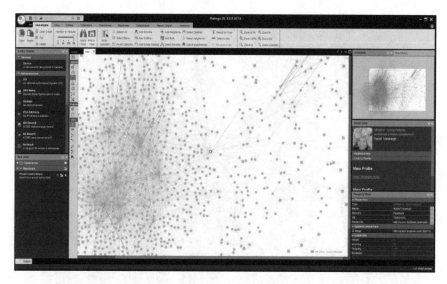

Fig. 7 A node-based graph suited for performing link analysis

Maltego can determine the relationships between the following things:

a. People:

- Names.
- Email addresses.
- Aliases.

b. Social networks (groups of people).
c. Organizations.

- Websites.
- Domain.
- DNS names.
- Net blocks.
- IP addresses.

d. Affiliations.
e. Documents and files.

3.4 Metasploit

Metasploit [8] gives information about security susceptibilities, and it is mainly used as a tool for developing and executing exploit code against a remote target machine.

Fig. 8 Metasploit framework for exploitation and testing

- First, open the Metasploit Console in Kali. Then move to Applications → Exploitation Tools → Metasploit.
- Then, the following screen appears (Fig. 8). If we want to find out the exploits related to Microsoft, the command can be msf → search name: Microsoft type: exploit (Fig. 9), where search is the command, name denotes the name of the object that we are looking for, and type denotes the particular kind of script we are in search of.
- Module or platform provides the information regarding the author name, vulnerability reference, and the payload restriction.

This is an example of how to see exploits related to Microsoft, and there are many commands in Metasploit framework for exploitation and testing, which can be checked using help command.

3.5 *Armitage*

Armitage [9] GUI for Metasploit is an accompaniment tool for Metasploit. It envisions targets, recommends exploits, and exposes the advanced post-exploitation features (Fig. 10).

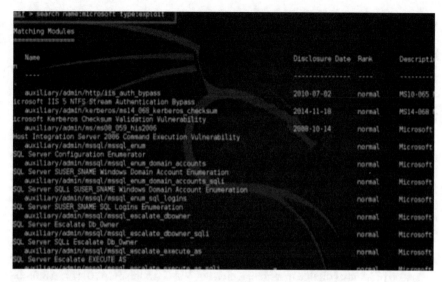

Fig. 9 Metasploit framework for exploitation and testing using help command

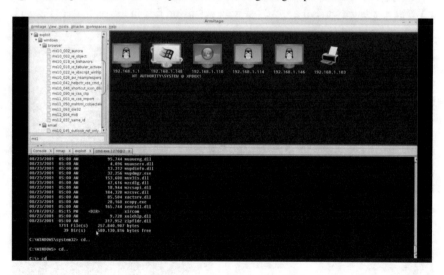

Fig. 10 An Armitage GUI for Metasploit

- Connect to Armitage, and it will list all the discovered machines to be exploited. Hacked target is shown in red colour with a storm with it.
- After having to target hack, just right-click on it and continue the exploration.

3.6 Hydra [10]

Login cracker tools that supports numerous protocols (Cisco auth, CVS, FTP, Cisco enable, HTTP(S)-FORM-GET, HTTP(S)-FORM-POST, HTTP(S)- GET, HTTP(S)-HEAD, HTTP-Proxy, ICQ, IMAP, IRC, LDAP, MS-SQL, MySQL, NNTP, Oracle SID, PC Anywhere, PC-NFS, POP3, PostgreSQL, RDP, Rexec, Rlogin, Rsh, SIP, SMB(NT), SMTP, SMTP Enum, SOCKS5, SSH (v1 and v2), SSHKEY, Subversion, Telnet, VMware-Auth, VNC, and XMPP) to attack.

- It will open terminal console (Fig. 11). In this case, we will brute force FTP service of Metasploitable machine, which has MAC Address CA: 01:17: A8:00:08 (Fig. 12).
- We have created in Kali a word list with extension first in the path user (Fig 13).

Fig. 11 Hacking router password using HYDRA

Fig. 12 Hacking MAC address using HYDRA

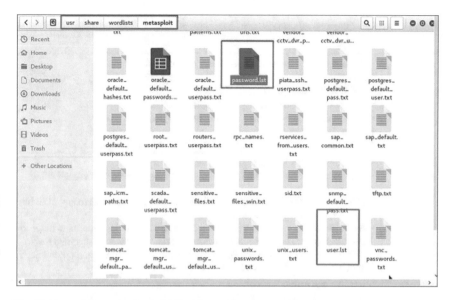

Fig. 13 A word list generation in Kali with extension

- The command is as follows: hydra-l/user/share/word lists/Metasploit/user-P/user/share/word lists/Metasploit/passwords ftp://192.168.2.58 V where V is the user name and password while trying (Fig. 14 and Table 1).
- As shown in the following, the user name and password are new local admin and $uP3r5ekrItpass, respectively.

```
msf auxiliary(smb_enum_gpp) > run
[*] 192.168.2.58:445    · Connecting to the server...
[*] 192.168.2.58:445    · Mounting the remote share \\192.168.2.58\SYSVOL'...
[+] 192.168.2.58:445    · Found Policy Share on 192.168.2.58
[*] 192.168.2.58:445    · Parsing file: \\192.168.2.58\SYSVOL\pwnlab.lcl\Policies\{31B2F340-016D-11D2-945F-00C04FB984F9}
\MACHINE\Preferences\Groups\Groups.xml
[+] 192.168.2.58:445    · Group Policy Credential Info
=============================

Name                    Value
....                    .....
TYPE                    Groups.xml
USERNAME                new_local_admin
PASSWORD                $uP3r5ekrItpass
DOMAIN CONTROLLER       192.168.2.58
DOMAIN                  pwnlab.lcl
CHANGED                 2016-07-12 07:04:23
NEVER_EXPIRES?          0
DISABLED                0

[*] 192.168.2.58:445    · XML file saved to: /opt/metasploit/apps/pro/loot/20160712000840_default_192.168.2.58_windows.g
pp.xml_841625.txt
[+] 192.168.2.58:445    · Groups.xml saved as: /opt/metasploit/apps/pro/loot/20160712000840_default_192.168.2.58_smb.sha
res.file_786986.xml
[*] Scanned 1 of 1 hosts (100% complete)
[*] Auxiliary module execution completed
```

Fig. 14 Track username and password

Table 1 Features of the tools used are shown as follows

Tools	Features
Nmap	• It is a network scanner tool which is free and open source • It is connected to a network/host and discovers open ports (responding to TCP and ICMP requests) • It points out hosts answering to network requests • It gives the host details (like network and OS details) • It can discover application running and its details • It is used to scan huge networks consisting of thousands of machines
Burp Suite	• It is a Java-based Web penetration testing framework • It helps to identify vulnerabilities and verify attack vectors that are affecting Web applications • It acts as 'man in the middle', capturing and analysing requests to and from the target Web application to be analysed • It can be paused, manipulated, and replay individual HTTP requests in order to analyse potential parameters or injection points. Those can be specified for manual as well as automated fuzzing attacks to discover potentially unintended application behaviours, crashes, and error messages
Maltego	• It is generally used for open-source intelligence and forensics • It helps to discover data in visual formats using built-in libraries of transforms • It permits creating custom entities apart from the entities that it provides • It analyses social, computer, or any real-world relationships from data sources, DNS records, search engines, API, social network, and various meta-data
Metasploit	• It is used to check security vulnerabilities and also for penetration testing IDS signature development • It includes anti-forensic and evasion tools, which are built on its framework • To choose an exploit and payload, some information about the target system is needed, such as operating system version and installed network services • It can be gleaned with port scanning and OS fingerprint tools such as Nmap • It can import vulnerability scanner data and compare the identified vulnerabilities to existing exploit modules for accurate exploitation
Armitage	• It is a graphical cyber-attack management tool under Metasploit framework • It visualizes the potential exploits and recommends too • It is used to access the advanced features of Metasploit • A user may launch scans and exploits, get exploit recommendations, and use the advanced features of the Metasploit Framework's metaoperator
Hydra	• It includes many login protocols like FTP, SMB, POP3, IMAP, VNC, and SSH • It is known as paralyzed network logon cracker • It is used for brute-force attack on any protocol like password and username guessing if any field is provided, cracking login credentials, etc. • A very well-known and respected network logon cracker (password cracking tool) which can support many different services

4 Advantages of Ethical Hacking

Millions of systems are hacked every second for the monetary as well as economic benefits resulting in a slowdown in the growth of a country. Hacking is a process which requires high profile techniques to catch the data theft and fraud. These techniques may or may not be within the permissions of the cyber laws. Benefits of ethical hacking are noticeable, but many are overlooked. The benefits include:

- National security breaches and fighting against terrorism.
- To prevent malicious hackers from gaining access to the computer system.
- Having acceptable pre-emptive measures in place to prevent security breaches.

5 Limitations of Ethical Hacking

An ethical hacker should know the consequences of illegal hacking into a system. Ethical hacking is usually conducted in a systematized manner, usually as part of a penetration test or security audit. The ethical hacker uses the knowledge they have when they are involved in malevolent hacking activities. This also referred to as intrusion testing, penetration testing, and red teaming. The ethical hacker generally sends and place spiteful code, viruses, malware, and other harmful things on a computer system.

6 Conclusion

Information is the most valuable strength of any organization. Hacking is the activity through which intruders are trying to gain access to the system to steal personal/corporate data. Everyone should pay much attention so that security measures can be strong to protect the confidentiality and integrity. Ethical hacking identifies and rectifies weaknesses of the computer system by describing the process of hacking in a decent manner. It protects the privacy of the organization and informs the hardware and software vendors of the identified weakness. Ethical hacking thus improves the security of the network or system.

References

1. Kumar, D., Agarwal, A., Bhardwaj, A.: Ethical hacking. Int. J. Eng. Comput. Sci., 4(4), 11466–11468 (2015)
2. Sahare, B., Naik, A., Khandey, S.: Study of ethical hacking. Int. J. Comput. Sci. Trends Technol. (IJCST) 2(4) (2014)

3. Munjal, M.N.: Ethical hacking: an impact on society, cyber times. Int. J. Technol. Manag. **7**(1) (2014)
4. Utkarsh, K.: System security and Ethical hacking. Int. J. Res. Eng. Adv. Technol. (IJREAT) **1**(1) (2013)
5. Juneja, G.K.: Ethical hacking: a technique to enhance information security. Int. J. Innov. Res. Sci., Eng. Technol. **2**(12) (2013)
6. Tekade, A.P., Gurjar, P., Ingle, P.R., Meshram, B.B.: Ethical hacking in linux environment. Int. J. Eng. Res. Appl. (IJERA) **3**(1), 1854–1860 (2013). ISSN: 2248-9622
7. Begum, S., Kumar, S.: Ashhar: a comprehensive study on ethical hacking. Int. J. Eng. Sci. Res. Tecgnology, **3** (2016). ISSN: 2277-9655
8. Ajinkya, A.F., Kashikar, A.G., Zunzunwala, A.: Ethical hacking. Int. J. Comput. Appl. (0975–8887), **1**(10), 14–20 (2010)
9. Whitman, M.E., Mattord, Herbert, J.: Management of Information Security, Boston, Massachusetts: Thomson Course Technology, pp. 363–375 (2004)
10. Smith, B., Yurcik, W., Doss, D.: Ethical hacking: the security justification. In: Proceedings of the Ethics of Electronic Information in the 21st Century Symposium (EEI21), Inc. Publishers, University of Memphis, Memphis TN USA (2001)
11. Satapathy, S., Patra, R.R.: Ethical hacking. Int. J. Sci. Res. Publ. **5**(6) (2015)
12. Mukhopadhyay, R., Nath, A.: Ethical hacking: scope and challenges in 21st century. Int. J. Innov. Res. Adv. Eng. (IJIRAE) **1**(11) (2014). ISSN: 2349-2163

Internal Organizational Security

Rehan Aziz, Sohang Sengupta and Avijit Bose

Abstract The triad of CIA, i.e., confidentiality, integrity and availability, are the key principles that are essential for any security system. It deals with confidentiality, which makes sure that no unauthorized person can access the system, integrity, which keeps the track of data such that it does not change in any manner and that its authenticity is maintained, and availability, such that it is readily accessible to those it is intended to. The most endangered areas of a given system are the location of the server and the places from where it can be accessed, the virtually secured region of an organization like the workplace where all the data available to its employees are kept and the last is the mind of the employees which, however limited, can be crucial. It may happen that an employee having limited data at his/her disposal becomes a major security threat. The tally and threat of the attacks of this nature are increasing with every new security measure being placed into action. The attackers are deploying new methods to break into the institutional structure. Therefore, it is necessary to look into all the possible threats so that the attacker has very limited to no access to the internal physical infrastructure. The aim of the following paper is to help new and existing organizations in developing a physical security system which will look into the internal security from all angles.

Keywords Surveillance · Internal threat · Code of conduct

1 Introduction

In a survey done by 'Cybersecurity Insiders' titled 'Insider Threat 2018 Report' [1], it was found that 'two-thirds of organizations (66%) consider malicious insider attacks or accidental breaches more likely than external attacks.' So when an organization prepares its annual plan for security, it is important that extensive care should be taken while dealing with the internal threat. If we look into the number of attacks, at least 50% experienced some internal attack, with 12% experiencing more than 10 attacks.

R. Aziz (✉) · S. Sengupta · A. Bose
Institute of Engineering and Management, Kolkata, India
e-mail: rehan4aziz@gmail.com

© Springer Nature Singapore Pte Ltd. 2020
M. Chakraborty et al. (eds.), *Proceedings of International Ethical Hacking Conference 2019*, Advances in Intelligent Systems and Computing 1065, https://doi.org/10.1007/978-981-15-0361-0_17

This is highly awakening as these attacks are carried out by individuals who are a part of the organization and also the fact that these figures are rising continuously. It not just exposes the company to various security threats but also has a direct effect on the financial status and public image. So in this paper, we will look into various aspects when it comes to internal security with extra focus on employee surveillance and the importance of code of conduct.

1.1 Surveillance Method

As Justice Mathew put it, 'The right to privacy will, therefore, necessarily, have to go through a process of case by case development' [2] (Govind v. State of Madhya Pradesh, AIR 1975 SC 1378). The given observation will hold for each and every instance where the term 'privacy' is applied. Be it the case of individual versus government or in the current context of workplace surveillance versus an employee's right to privacy.

If we take an overview of implementation of privacy laws in our country, in the report prepared by the planning commission [3], two contrasting cases are discussed where the observation of the first case was 'the right of privacy is not a guaranteed right under our constitution, and therefore the attempt to ascertain the movements of an individual is merely a manner in which privacy is invaded and is not an infringement of a fundamental right guaranteed in Part III.' And the observation of latter one was 'rights and freedoms of citizens are set forth in the constitution in order to guarantee that the individual, his personality and those things stamped with his personality shall be free from official interference except where a reasonable basis for intrusion exists' even though both the cases share similar appeals. It may be an out of context case, but it does give us a widespread picture of how the privacy laws are implemented. So one has to ask the very question, how far are we allowed to go when it comes to protecting our interests? There cannot be and, in our country, is not any single law which governs the power that an organization has over its employee. Many a times, a constitutional provision is weighted against another to find that in the current context which should be upheld.

From various landmark judgments like the USA vs Hamilton [4], it can be said that it is reasonable that any activity done in companies' device is subjected to surveillance and does not interfere with an individual's fundamental right. In another case, Bărbulescu v Romania [5], it can be noted that the employee was fired after the company intercepted some personal email he had sent, the court upheld the decision, and only one of the judges presiding over the case was opposed to the verdict, stating that employers must make their rules regarding Internet usage clear to employees. Various similar cases show that monitoring of online activities of employees, like on social media platforms, is allowed as long as clear instructions have been laid down regarding it. Apart from online activity surveillance, camera monitoring is allowed and background check is a common policy. Further surveillance is always subjected to the relevance.

1.2   Hierarchy and Related Threats

When we talk of data and information security in the workplace, it is necessary to look into the different sets of people available at an organization and also a different nature of threats each set possesses. A different set of people poses a different set of threats according to accessibility to the data and information of the organization and also the nature of them.

If we will look into the hierarchy of an organization, we will get the availability of information at different levels. The highest level of information or data is available with the board or the president itself. The level of threat there is negligible as they are the people with personal interest directly linked to the organization itself, and so any harm to the organization will result in their personal loss.

Next comes the VP with the most information and with accessibility to nearly all sensitive data. They possess a great threat to an organization if gone rogue, but these are countered by the fact that a person is generally given such a role after being a part of the organization for a long period, so attributes like loyalty counter the threat. Also, as these people are smaller in number, so the seniors take a direct interest in their day-to-day activities. So if the given individual gets in any undesirable situation, it would be solved with personal interest from the top. Nonetheless, a small level of surveillance would be more than enough to counter this level of threat.

Next come the department heads. This is a crucial level, as even though they do not have data of the complete organization, they have complete control of the information flowing through their department. For example, the sales head of an organization can sell their sales technique. The things stopping them from doing things which are not in the interest of the organization, apart from obvious reasons like morality, are that they are the ones answerable for any wrongdoing in their department and also like vps they have a personal interest to the organization as most of them are a longtime employee.

Next are the employees. Even though the information available at this stage is much less compared to other levels, they can be easily persuaded into getting involved in activities undesirable for the organization. There are quite a few reasons why someone at this level will go rogue. It may be a better pay raise and can be extortion as people here generally have less influence to counter it on their own. There are various other reasons.

Then comes the support staff. They are often neglected as they do not have access to any confidential data, but they do have full information of an employee's day-to-day life. They know everyone's behavior, their routines, their social networks, or in short they are the best surveillance tools for anyone trying to infiltrate an organization. They can easily be persuaded to do a job. Any information can be extracted from them without even them noticing.

We also have to look into a class of IT professionals. As recent studies [1] have shown that apart from regular employees, IT people are the biggest security risks for the organization. The second and third levels, i.e., vps and department heads, are referred to as managerial posts, as they are the ones managing the day-to-day work of the organization. Surveillance in this category poses a different problem, that of prestige as these are generally high-end jobs and people involved here are generally long-term employees as well as have higher respect in the society. The last two sets, i.e., the employees and the support staffs, need to be severely monitored with extra focus on employees. But the problem arises when the surveillance methods interfere with an individual's privacy. So, it is necessary to find just the right balance.

1.3 Summary

While drafting a conclusion on how the security practices should be, one has to be ready to face many more cases like the above-mentioned one. One should know that even small things like the style of writing of the organization's law can be a big game changer when an unwanted lawsuit comes. One can always argue that a certain provision is applied so as the interest of the organization is upheld, but has to make sure that it is in no way legally or ethically wrong.

So coming back to the problem at hand, it can be dealt with proper interaction of two separate groups. The first one is IT professionals and the management which will decide on the extent of surveillance to be done on employees and various other methods applied which are discussed in Sect. 2. The other one is the Human Rights Department which with the help of the code of conduct will look over them so as the interest of the organization is not compromised in any way.

1.4 Other Threats

When we are looking into the possible threats, it is necessary to look into the reasons why an individual will be involved in an activity which is not only harming the interest of the organization they are a part of but at the same time illegal.

Such activity can only be brought about by a few reasons. One is that the person is forced by someone or something which is not in their power, or it could be personal enmity or just a notorious activity. If it is the case of unwanted wrongdoing, some consideration can be taken while dealing with it as the said person does not have the intention but at the same time is ethically and legally wrong. If such an activity is carried by a trained professional, it would be even difficult to detect with the surveillance means at hand. Generally, in these cases, the person would join the organization with the wrongful intent. It is the job of the hiring staff, or as in most of the cases, the HR Department to avoid such an individual or mark them so as extra precaution could be taken while dealing with them. Many times, it becomes unavoidable. In that case, one can only depend on a better secured internal infrastructure, which is discussed in Sect. 2.

Then, there comes a category where security lapse occurs due to accident and/or negligence of the person involved. According to a survey [1], around half of the internal security threat is because of these errors. Upgrading the security according to this case is quite a hectic task, as the person in question is generally unaware of the consequence of the mistake. The problem will solve itself if the employees are properly trained to handle the infrastructure available to them. Therefore, proper hiring of the staff, regular workshops and training regarding the new and upcoming technologies are essential.

2 Safety Measures

When it comes to surveillance, we have to first establish who the target is and whom it is to be reported to. When it comes to monitoring, will there be a completely autonomous algorithm at work which will do uniform surveillance of every member of the organization, or will there be an algorithm monitored by a set of people dedicated to keep an eye on the people working there, be in the form of HR Department or the IT Department.

2.1 Machine and Human Error

Having a dedicated team to carry on the surveillance process would mean dealing with an unavoidable human error. It may be in the form of mismanagement of personal data, negligence of duty or addressing of wrong suspect.

Employee surveillance in the picture of user entity behavior analytics: User entity behavior analytics when combined with machine learning develops a standard of normal human behavior by gathering information about an individual's daily lifestyle which includes parameters like social media presence, type of mail received, call frequency and similar things. It may be argued that a given baseline of behavior may not be common to every individual. In these cases, a different image can be developed for everyone under purview.

When implemented in an organization for the employee surveillance, it will raise an alarm when unusual behavior is detected. During the initial learning stages, there is a high probability that an alarm is raised even if the work is legitimate. That is the learning stage; we have to feed it to the system that it is normal behavior. Therefore, it can be argued that more the false alarm raised during the learning period, the better will be the accuracy after the initial learning period. It is also recommended to update the model regularly by inputting new data and classifying it as normal behavior or not.

2.2 Products Available to Increase Security

Data loss prevention: Data loss prevention, or DLP, is a set of technologies, products and techniques that are designed to stop any sensitive information from leaving an organization. Any data in the wrong hands can severely threaten the security and damage the infrastructure. It is generally a set of rules which detect unwanted data transfers or sensitive information flowing through the electronic medium.

Identity and access management: It is a framework of policies and technologies for ensuring that the proper people in an enterprise have the appropriate access to technology resources [6]. It identifies, authenticates and authorizes individuals who will be utilizing IT resources, hardware and applications employees need to access. It addresses the need to ensure that only the person authorized accesses the resources of the organization.

Security information and event management (SIEM): SIEM technology supports threat detection and security incident response through the real-time collection and historical analysis of security events from a wide variety of event and contextual data sources [7]. It also investigates incidents and does compliance reporting by analyzing existing data from these sources. The main work of SIEM technology is event collection and the ability to correlate and analyze events across different sources.

Cloud access security brokers (CASBs): CASBs are on premises, or cloud-based security policy enforcement points, placed between cloud service consumers and cloud service providers to combine and interject enterprise security policies as the cloud-based resources are accessed [8]. CASBs consolidate multiple types of security policy enforcement. Example: Security policies include authentication, single sign-in from an account, authorization, logging malware detection/prevention alerting, encryption, tokenization and so on.

3   Code of Conduct

3.1   Importance

A code of conduct is the set of guidelines provided to the employees so one can look over it to keep their day-to-day conduct within the rules and regulations of the organization. A well-written code of conduct is a salient feature of any company's policy as it affirms its objectives and principles, and at the same time disclosing what values they expect from leadership as well as their whole workforce. It also guides them to the significant point that needs to be kept in mind while making important decisions that may affect the interest of the organization in any way.

Many a time, other organizations tend to look over the code of conduct of the institution they are getting in business with. It gives them a broader perspective of the schemes of others, meanwhile giving them an overview of the individuals they will be working with. Accordingly, they will be making their plan of action.

For the employees of the firm, it will be a rule they have to follow to avoid any consequences or in some cases termination, because of failure to abide by the companies' policy. Furthermore, acknowledgment of the ethical part of the code will to some extent keep them from getting involved in any activity which is not in the interest of their working area, or as in ongoing discussion prevent them from being rogue.

So here, we will look into some of the points given by ECI [9] in detail that should be exclusively added into the code of conduct, apart from the usual ones, to avoid the case discussed above.

3.2 Principal Points

Handling of company data and information

- Maintaining record—any flow of data through an individual should be recorded in a personal capacity, or reported, as per the policy.
- Privacy and confidentiality—generally, there are clear instructions on which data have to be confidential and are out of bound.
- Disclosure of information—there is some information which has to be made public or disclosed to a given set of individuals to maintain transparency.

Conflict of interest

- Disclosure of financial interest—an important disclosure, as any unaccountable growth may refer to individuals' involvement in unwanted activities.
- Gifts and gratuities—any gifts received in professional capacity need to be reported, or their record to be kept.
- Outside employment—an important point, any outsider has a comparatively high chance of a conflict of interest with the place of employment as it puts them in a security risk.

Ethics and compliance resources. This is one of the most important points when it comes to ethics as it states the protection available for those who report the wrong-doings as well as consequence for the wrongdoing.

- Ethics advice helpline
- Reporting procedures
- Anonymous reporting helpline
- Summary of the investigation procedure
- Anti-retaliation policy and protection for reporters
- Accountability and discipline for violators.

Internet and social media

- Use of Internet using company devices—accessing social media through company devices may put them under the scanner, so it is generally recommended not to access personal accounts with it.
- Blocked sites—some contents are blocked by the organizations, or it is recommended not to access them due to various reasons.
- Posts about the organization—generally any and all the posts regarding the organization are kept under surveillance even if it is in personal capacity.

3.3 Summary

A code of conduct is not the ultimate resource when it comes to dealing with the problem at hand, but it is a resource that will at least reduce the risk to a significantly small number. Many individuals will be forced to rethink their daily activities. It does not mean that there will be any restraint, but they will be prioritizing the organization's interest over others. As mentioned in Sect. 2.2, the last two steps of hierarchy, i.e., employees and support staff, will be most influenced by the code, but the managerial post will also look into it as whatever the limited threat they pose can be countered. They will be more careful in taking any important decisions as not only their job, but their career as a whole depends on it and also all the respect they have gained over the years.

4 Conclusion

The aim of the paper is to find a way to prevent and detect any rogue element which may be present in an organization. To find that, we have to get the right balance between discussions in Sects. 2 and 3.

The first step while securing is to develop a system that will prevent any individual from getting into a place from where they can access confidential information and carry on an attack. The next step is detecting the attack and stopping it before any valuable data are leaked. This also involves segregating data in such a way that even if someone can get access to some data, other valuable ones are secured, i.e., minimize the data leakage. The last thing is damage control, i.e., what should be done after an attack took place.

The first step is to investigate how the attack took place, was an alarm raised, have IDS (intrusion detection system) detected anything or was it simply a false positive. The next step is determining the nature of the attack, was it attempted ransomware, was it an attempt to hack into the internal organizational architecture, was it a [10] spear-specific file or if there is any other reason involved. The next step is auditing the security practices in place. A detailed report should be made for it. The last step is upgrading the security practices so any future attacks can be avoided.

So, in conclusion, a good surveillance system supported by different security practices, a dedicated team that can stop any ongoing attack and a very strong incident response team with a tried and tested damage control in place can minimize the risk pertaining to any organization.

References

1. Insider Threat, 2018 Report, Reported by CA Technologies
2. Govind v. State of Madhya Pradesh, AIR 1975 SC 1378
3. Report of the Group of Experts on Privacy, Planning Commission (Chaired by Justice A P Shah, Former Chief Justice, Delhi High Court)
4. EPIC, United States vs Hamilton. https://epic.org/amicus/hamilton/
5. Tech2, First Post. https://www.firstpost.com/tech/news-analysis/right-to-privacy-we-take-a-closer-look-at-employer-employee-safeguards-in-india-4015201.html
6. Wikipedia. https://en.wikipedia.org/wiki/Identity_management#cite_note-:0-3
7. Gartner.   https://www.gartner.com/it-glossary/security-information-and-event-management-siem
8. Gartner. https://www.gartner.com/it-glossary/cloud-access-security-brokers-casbs/
9. ECI. https://www.ethics.org/resources/free-toolkit/code-provisions/
10. Norton Blog: https://uk.norton.com/norton-blog/2016/12/what_is_spear_phishi.html

Securing Air-Gapped Systems

**Susmit Sarkar, Aishika Chakraborty, Aveek Saha, Anushka Bannerjee
and Avijit Bose**

Abstract A security measure which involves isolation of a computer or a network in order to prevent it from establishing an external connection, such a security measure is called air gapping. If we are able to physically segregate a computer in such a way that it becomes incapable of connecting wirelessly or physically with other computers network or devices, then that computer is an example of an air-gapped computer. Since air-gapped computers are neither connected to the Internet nor other networks that are connected to the Internet, this makes hacking such computers quite difficult. If we are able to cut off any connection to the computer, then only we can guarantee that no third party would be able to access the client. The object of this paper is to verify the feasibility of air gapping.

Keywords Air gapping · Covert channels · Security · HVAC · Attack model · Data exploitation

1 Introduction

A true air gap denotes that the computer or the network must be physically separated from the Internet and data transfer can occur only with the help of USB flash drives, other removable media or even a firewall which directly connects two computers. However, there are a lot of companies which consider a network or system as sufficiently air gapped even if it is isolated from other computers or networks through a software firewall. Though the presence of security holes or even if the firewalls is configured insecurely they can be breached. Earlier, we used to believe that air-gapped systems require an attacker to have physical access to breach them. However,

S. Sarkar (✉)
Department of Computer Science Engineering, Institute of Engineering & Management,
Kolkata, India
e-mail: susmit.sarkar97@yahoo.com

A. Chakraborty · A. Saha · A. Bannerjee · A. Bose
Department of Information Technology, Institute of Engineering & Management,
Kolkata, India

© Springer Nature Singapore Pte Ltd. 2020
M. Chakraborty et al. (eds.), *Proceedings of International
Ethical Hacking Conference 2019*, Advances in Intelligent
Systems and Computing 1065, https://doi.org/10.1007/978-981-15-0361-0_18

this strategy is not completely secure or foolproof. Attackers have developed several models to infiltrate the gap between these networks and hence attack the secure systems.

One of the most well-known attacks occurred when the virus "Stuxnet" was used to target the Iranian nuclear program. The virus was used to intrude upon a computer network that was isolated as it controlled one of the centrifuges of the nuclear power plant and hence effectively causing a meltdown. Another example of such an attack would be "agent.btz", which is essentially a malware that was targeted against the USA military. This was the first effective breach of USA military sensitive assets which were air gapped for enhanced security.

2 Related Works

The hidden channels are usually discussed in the professional literature using several methods [1–3]. On the other hand, in this document, we focus on secret channels that can connect laptops, or physically separate. This subgroup in the hidden channels uses some physical phenomena, using inaudible audio channels, optical channels and electromagnetic emissions. Madhavapeddy [4] and others argue over audio networks, while Hanspach and Goetz [5] discover the secret method of networking near ultrasound. Loughry and Umphress [6] discuss the flow of information optical secretions. Transfer of hidden data based on data, use of electromagnetic emission, discussed by Kuhn and Anderson [7]. Guri et al. [8] feature "AirHopper", malware using FM emissions to clean the data of an airplane. Note that the channels mentioned above are and they do not allow you to create an input channel in an isolated network. Technically, thermal radiation is a form of electromagnetic emanation; in fact, the old Murdoch was proposed [9] to be used as a hidden channel.

The general topic of hidden channels used by malware was intensively studied for more than twenty years. In order to avoid the detection of firewalls and IDS and IPS systems, attackers can hide leaked data on legitimate Internet traffic. Many protocols have been studied over the years in the context of the desired channels, including TCP/IP, HTTPS, SMTP, VOIP, DNS queries and others [10]. Other types of hidden channels include a time channel where the attacker encodes data using packet synchronisation [11] and image steganography [12], where the attacker inserts data existing image. The hidden channel subproject focuses on leakage of hidden data from airless computers, where Internet connection is not available to the attacker. Air opening hidden channels that can be classified as electromagnetic, the subject was acoustic, thermal and optical channels scientific research for twenty years.

In hidden electromagnetic channels, electromagnetic emissions generated by various hardware components below the computer are used to perform information leakage. Kuhn and Anderson [7] introduced the attack ("soft-tempest") with participation transfer of hidden data using electromagnetic waves coming from a video cable. Emissions are generated when the created images are sent to the screen. 2014, Guri et al. introduced AirHopper [8, 13], the type of malware to close the air gap between computers and a mobile phone nearby with FM radio signals from short. In 2015, Guri et al. presented GSM [14], malware it can generate electromagnetic emissions on cell frequencies (GSM, UMTS and LTE) in the PC memory bus. The study showed that modulated emission data can be selected by rootkit, which is in the baseband program mobile phone nearby. USBee [15], presented in 2016 by Guri et al. used a USB data bus to generate electromagnetic signals and modulate digital data using these signals.

Advanced Persistent Threats

An APT or advanced persistent threat is essentially a form of obfuscated computer network attack in which the attacker gains unauthorised access to the network, escalates the privileges of the system, and creates a persistent backdoor for leaking out sensitive data from the victimised system. The term "advanced" refers to the sophisticated exploitation techniques used, the "persistent" refers to the creation of the backdoor and the "threat" refers to the attacker or the process of attacking the system. The following section gives a deeper insight into the stages of an advanced persistent threat.

2.1 Attack Cycle

In almost every case having a persistent backdoor channel to the attacked system is essential, so that the attacker can derive information from the malware that the system has been infected with. The attacker can even modify the malware signature or make it metamorphose to prevent detection. Gaining persistent access occurs in three stages. In stage 1 is the reconnaissance, where the attacker researches about the system to be targeted in order to obtain as much data as possible so that the best fit pathway to intrusion can be determined. In stage 2 is the method of initial intrusion is accomplished via methods of social engineering, phishing or USB plug and play device. One of the targeted machines are infected then begins the next stage (Fig. 1).

In stage 3, the malware begins searching for sensitive data such as server, subsystem, database, etc. It breaks into these resources and sends the information back to the attacker or mutates the attack resource as per the code written into it. In the next stage, the malware begins execution of its code by replicating itself and infecting other host machines and in the last stage it deletes all its record logs and self destructs itself.

2.2 Proposed Exploitable Vulnerability

In this paper, we would like to propose a new model which can essentially exploit and enable a remote hacker to be able to send commands to hosts that are infected,

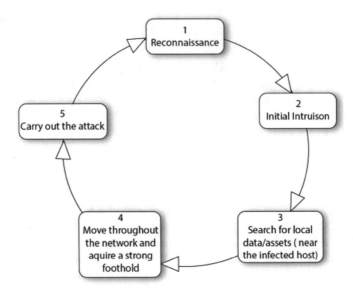

Fig. 1 Stages of an APT on a victim's network

even over the air gap. The attacker would have to penetrate a heating ventilation and air conditioning system (HVAC) network which is placed in the same network as the infected host and then modulate the air temperature to send signals to the malware inside the host. The infected host's internal thermal sensor picks up these thermal signals, which is relayed to the malware it is infected with and hence, the malware executes the commands as commanded. The purpose of this model is to establish a covert broadcast channel which is unidirectional to infected hosts within the air-gapped network. Such kind of a model can be used by the attacker to manage persistent attacks remotely and can also lead to initiate a denial of service (DoS) attack to interrupt daily work or to distract from an ongoing attack.

3 Attack Scenario

In this particular part, we have detailed out a feasible attack scenario in which our proposed model can be used to the hacker's advantage.

3.1 Attack Motivation

Assuming the target organisation is based on a contained Ethernet environment which is essentially disconnected from the World Wide Web and from any other

Table 1 Examples of attacks that can be performed on target network

Task	Description
Search and delete a file	The removal of sensitive files for the advantage of some cause. For instance, military intelligence documents or digital evidence that supports a certain case. This can be done by searching for keywords
Search and edit a file	The changing of files or data entries found by a keyword similarity search. For instance, replacing names of people or locations or other string swaps that violate a file's integrity
Temporarily disable a system	Disabling a host, server or subsystem by means of an internal DoS attack or the direct intervention of some infected host
Temporarily disable a security protocol	Disabling the security measures of a host, subsystem or even a security authentication waypoint for personnel
Move collected data to a staging area	In the case, where sensitive information has been gathered by the infected hosts, the data can be copied to a common extraction point. For instance, an insider whose presence within the isolated network is only temporary
Self-destruct	Covering tracks. All evidence of the infection and log files that may indicate the existence of the APT are deleted
Encryption key change	The changing of the encryption key is used in the communication between infected hosts or their encrypted log files

physical LAN networks. Nevertheless, the encompassing building in parallel to the organisation's network and it is a HVAC system which is connected to the Internet.

In this particular scenario, if the attacker intends to perform an advanced persistent threat attack on the organisation, he attacks the random hosts behind the air gap using any kind of social engineering or phishing attacks.

The attacker in this scenario needs flexibility to control the attack by activating the operations in Table 1 when necessary. For example, the attacker may want to search and modify the file online several years later (other examples can be found in Table 1). To achieve this, the attacker needs the ability to cover commands to prevent infected machines from being left behind the air space. Therefore, the attacker plans this forward and uses the resistance model in this article.

3.2 Proposed Model of Attack

A general description of the proposed model of an attack is visible in the use case (Fig. 2) and is performed as follows. Step 1: Attacker attacks are sent to the command

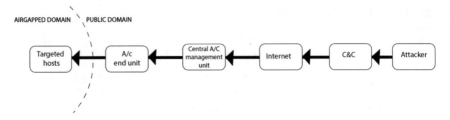

Fig. 2 Use case of the proposed attack model

and control server on the Internet. Add the degree of separation between the attacker and the attack to protect the identity of the attacker. Steps 2–3: The server sends commands through the system's online Web interface to the central management unit of the infected HVAC system (for legitimate remote control of the HVAC system). Steps 4–5: The HVAC system transmits commands through the air by changing the heat environment according to a predefined protocol.

This attack model can be called an APT attack in the following way. In the time of the reconnaissance stages, the attacker determines the type of HVAC and its protocols that are running in the building management. After gaining control over the HVAC system, the attacker then creates a malware client which is tasked to receive and interpret the thermal changes in the form of signals from the environment.

Since most computers use thermal sensors to activate cooling systems inside the CPU, these client-side receiver systems are not difficult to build and can use the BIOS's in built interpreter to interpret these signals. Once the malware takes over the control from the host, it can exploit these sensors in turn to sense the changes in the temperature of the surrounding environment.

After the malware has completed interpreting these signals, it then goes on to infect the targeted host networks using the methods discussed in the attack cycle. The last stage of the attack comes in as the malware replicates itself and attacks other neighbouring systems, autonomously gaining various assets on its way.

At this particular point, the attacker has a unidirectional covert path of signal exchange from the malware-infected source to the attackers system by the infected HVAC system and can continuously issue commands to the malware to make the necessary changes and to deploy the designated attacks that the attacker wishes to impose. Moreover, now that attacker can also perform a DoS or denial of service attacks to create a distraction while the malware completes its designated task of corrupting or leaking information.

4   Experimental Results

In this particular section, we have presented the results of the experiments to prove the concreteness of the formation of a covert thermal channel from an HVAC system to a nearby computer is possible. The results show the changes caused by the infected

Table 2 Air conditioning unit's impact on the motherboard's sensor

The air conditioning unit's impact on the motherboard's sensor after noise mitigation	
Lowest recorded temperature (°C)	22
Highest recorded temperature (°C)	30
Maximum temperature difference (°C)	8
Distance from transmitter (m)	3
Dimensions of the office room (m³)	$4 \times 4 \times 5$
Linear rate of ascent (°C/min)	1.23
Linear rate of descent (°C/min)	−1.24

HVAC system in the room temperature, that are detectable by a common desktop computer and that can easily be converted into signal by any form of line encoding techniques.

To test our hypothesis, we created a probable frame which an attacker might broadcast to all receiving computers within the same HVAC system. The frame comprises a "Change Encryption Key" which is used for internal communication which is an OP Code and a payload of 128 bit key, and therefore, we assumed that internal cooperation is present between infected hosts and therefore on receiving this command all infected hosts will update their channel encryption keys.

We have found that the motherboard thermal sensor had a faster response to the changes in the external environment and therefore we have used the motherboard's thermal sensor for testing the model. On analysing the first step of the response from the infected air conditioner, we see that there is a dip in temperature from H to L on the sensor, i.e. 26–23 °C (Table 2). Therefore, we decided to test the model for 1.5 min and hence calibrated the other parameters accordingly (Table 3). In this particular setting, we could transmit a frame of 134 bits within three hours and thirty

Table 3 Parameters taken for testing the attack scenario

Parameters taken for testing the protocol	
Parameter	Value
C	Small office room
T	Centralised AC system
R	Closed chassis desktop computer
H	26 °C
L	23 °C
D	3 °C
γ	1 °C
μ	0.01 °C
MAF window	1 min of samples
Sample rate	3.3 Hz

minutes with a constant bit rate of 40 bph and then again a frame of 134 bits but this time the receiver was sampling the inputs from the sensor and demodulating the bits.

Therefore, we came to the conclusion that 40 bph is more than enough for any hacker to be able to trigger any task from the list given in Table 1 and hence we conclude that our attack model is actually feasible.

5   Forensics and Countermeasures

There are numerous ways in which the thermal convert channel attack model opposed in this paper can be detected and even prevented. These are represented in the attack model diagram Fig. 2 shown above.

5.1   Security Through Obfuscation of HVAC Web Interfaces

The most direct way of preventing this attack is to increase security in the HVAC systems. These systems should not be allowed to have direct internet access, and if that is necessary then it should be protected by a strong firewall. Moreover, the maintenance personnel incharge of these HVAC devices should be made cyber threat aware and should be trained to not divulge their passwords. They should also be trained in deleting irregular and anomalous activities in the network which can lead to a possible breach and should be instructed to report any such activities immediately.

5.2   Monitoring Symptoms of Possible Backdoors

If disconnecting the HVAC systems from the Internet portal is not possible, then there is another way of detecting covert thermal channels. This can be done by placing thermal sensors in the room of the air-gapped systems. In the case of abrupt and sudden sharp changes in the room temperature, these sensors can pick up and log the changes and hence can send a report of suspicious activity as it might signify the presence of covert thermal channels in the room. However, this theory is yet to put into use and the significant development of an effective algorithm to detect and analyse these changes in temperature and needs to be worked on.

5.3   Searching for Malware Signatures

For an infected host, in order to detect and receive commands over covert thermal channels, it is required to continuously sample the thermal sensor readings. Then, it is expected to have any noise mitigation process to be processed on the samples. This particular kind of behaviour can be detected with the help of dynamic analysis of the code. This kind of signature will essentially involve access of high frequency to the thermal sensor APIs of the operating system. If by the dynamic analysis of the code, such kinds of signatures are found then it can be confirmed that there is a presence of covert thermal channels.

6 Conclusion

Therefore, we see that the popular method of enhancing security of networks by separating them or rather having a gap of air between them is not secure enough and is susceptible to outside attacks. In this paper, we have shown an attack model which could be used by an attacker to breach into the air-gapped networks by the low-security HVAC systems. We have also put forward ways of prevention and detection of this attack model and these measures can be carried onto other attack models as well. Hence, we conclude that air gapping of systems does not ensure foolproof security and also acknowledge the necessity of further research on the security of air-gapped systems.

References

1. Wu, Z., Xu, Z., Wang, H.: Whispers in the hyper-space: highspeed covert channel attacks in the cloud. In: USENIX Security Symposium (2012)
2. Jie Chen, G.V.: CC-hunter: uncovering covert timing channels on shared processor hardware. In: MI-CRO-47 Proceedings of the 47th Annual IEEE/ACM International Symposium on Microarchitecture (2014)
3. Lee, K. S., Wang, H., Weatherspoon, H.: PHY covert channels: can you see the idles? In: 11th USENIX Symposium on Networked Systems Design and Implementation (NSDI '14), Seattle (2014)
4. Madhavapeddy, A., Sharp, R., Scott, D., Tse, A.: Audio networking: the forgotten wireless technology. IEEE Perv. Comput. **4**(3) (2008)
5. Hanspach, M., Goetz, M.: On Covert Acoustical Mesh Networks in Air. arXiv preprint arXiv: 1406.1213 (2014)
6. Loughry, J., Umphress, A.D.: Information leakage from optical emanations. ACM Trans. Inf. Syst. Secur. (TISSEC) **5**(3), 262–289 (2002)
7. Kuhn, M.G., Anderson, R.J.: Soft tempest: hidden data transmission using electromagnetic emanations. In: Information Hiding. Springer, Heidelberg, pp. 124–142 (1998)
8. Guri, M., Kedma, G., Kachlon, A., Elovici, Y.: AirHopper: bridging the air-gap between isolated networks and mobile phones using radio frequencies. In: 9th IEEE International Conference on Malicious and Unwanted Software (MALCON 2014), Puerto Rico, Fajardo (2014)
9. Murdoch, S.J.: Hot or not: revealing hidden services by their clock skew. In: ACM Conference on Computer and Communications Security (2006)
10. Zander, S., Armitage, G., Branch, P.: A survey of covert channels and countermeasures in computer net-work protocols. IEEE Commun. Surv. Tutorials **9**(3), 44–57 (2007)

11. Cabuk, S., Brodley, C.E., Shields, C.: IP covert timing channels: design and detection. In: Proceedings of the 11th ACM Conference on Computer and Communications Security (2004)
12. Cheddad, A., Condell, J., Curran, K., Kevitt, P.M.: Digital image steganography: survey and analysis of current methods. Sig. Process. **90**(3), 727–752 (2010)
13. Guri, M., Monitz, M., Elovici, Y.: Bridging the air gap between isolated networks and mobile phones in a practical cyber-attack. ACM Trans. Intell. Syst. Technol. (TIST) **8**(4) (2017)
14. Guri, M., Kachlon, A., Hasson, O., Kedma, G., Mirsky, Y., Elovici, Y.: GSMem: data exfiltration from airgapped computers over GSM frequencies. In: 24th USENIX Security Symposium (USENIX Security 15), Washington, D.C. (2015)
15. Guri, M., Monitz, M., Elovici, Y.: USBee: Air-Gap Covert-Channel via Electromagnetic Emission from USB, no. arXiv:1608.08397 [cs.CR]

Data Security Techniques Based on DNA Encryption

Mousomi Roy, Shouvik Chakraborty, Kalyani Mali, Raja Swarnakar, Kushankur Ghosh, Arghasree Banerjee and Sankhadeep Chatterjee

Abstract Security of the digital data is one of the major concerns of the today's world. There are several methods for digital data security that can be found in the literature. Biological sequences have some features that make it worthy for the digital data security processes. In this work, DNA encryption and its different approaches are discussed to give a brief overview on the data security methods based on DNA encryption. This work can be highly beneficial for future research on DNA encryption and can be applied on different domains.

Keywords Digital data security · DNA encryption · Cryptography

---

M. Roy · S. Chakraborty · K. Mali
Department of Computer Science & Engineering, University of Kalyani, Kalyani,
Nadia, West Bengal, India
e-mail: iammouroy@gmail.com

S. Chakraborty
e-mail: shouvikchakraborty51@gmail.com

K. Mali
e-mail: kalyanimali1992@gmail.com

R. Swarnakar
Department of Computer Science, Kalyani Mahavidyalaya, Kalyani, Nadia, West Bengal, India
e-mail: rajaswarnakar1997@gmail.com

K. Ghosh · A. Banerjee · S. Chatterjee (✉)
Department of Computer Science & Engineering, University of Engineering & Management,
Kolkata, West Bengal, India
e-mail: chatterjeesankhadeep.cu@gmail.com

K. Ghosh
e-mail: kush1999.kg@gmail.com

A. Banerjee
e-mail: banerjeearghasree@gmail.com

© Springer Nature Singapore Pte Ltd. 2020
M. Chakraborty et al. (eds.), *Proceedings of International
Ethical Hacking Conference 2019*, Advances in Intelligent
Systems and Computing 1065, https://doi.org/10.1007/978-981-15-0361-0_19

1 Introduction

Modern generation cannot be imagined without the blessings of Internet. With the technological advancement, most of the data are transmitted over the network for different purposes and very low-cost communication is possible. This domain ever growing and load on the network is ever increasing. With the growing interest of Internet in this world, the most concerned thing is security of information.

Apart from various advantages, the data communication is suffering from the security threats that make the data vulnerable. Several persons are continuously trying to intercept the data traffic and steal sensitive information or hamper the integrity of the data. Everyday a lot of information flows around the globe through Internet [1]; at the same time, security threat is increasing exponentially. There are various adversaries or hackers who try to break integrity of data or steal the data. Nowadays, privacy or data breach evoked a lot of attention. It has havoc impact concerning individual or an organization. Everyone wants their data to be secured and strictly avoid unauthorized access. This motivated the emergence of Cryptography. Transforming a plain text into something unrecognizable by a human is a process known as cryptography [2]. Data security which is also termed as cryptography is most essential in data transmission. It is a challenge to provide complete data security. Various cryptography techniques and schemes have been proposed in the past decade and evolved day by day till the date such as DES, AES, and RSA. Conventional cryptographic techniques are not powerful enough to provide effective security of information or data [3]. So the security of the data is one of the prime objectives of the data communication and it is a necessary module of any kind of remote data communication systems. Several organizations communicate sensitive data that must be protected from any kind of unauthorized access. The technique of secret writing has been investigated from the ancient age. Modern-day data communication methods rely on cryptography, steganography, and some other techniques for the secured transmission of data [4, 5]. Cryptography focuses on the encryption of the data, and steganography deals with data hiding, i.e., hiding actual message in some other data. Both of the methods depend on complex mathematical operations for higher security. Different types of data with various features [6–8] make the encryption process challenging.

In recent years, another field of cryptography called DNA encryption is emerged and applied in several areas. In this approach, the actual data is hidden in the form of digital DNA codes. DNA encoding helps to preserve higher confidentiality compared to many traditional systems. It is believed that a huge amount of data can be efficiently stored using DNA encryption with better security. It is one of the powerful and widely used algorithms for data security [9].

DNA encryption is one of the emerging techniques in the area of data security in the twenty-first century. In this novel, scheme DNA used as information career and plain text is encoded in nucleotide sequence [10]. It is believed that DNA encryption provides higher confidentiality and can work with larger data due to its high storage capacity. One gram of DNA can contain 108 terabytes of data. DNA computing which is also known as molecular computing enhances the capability of parallel processing in molecular level, introducing a new data structure and a different calculation scheme [3]. The research in the field of DNA computation and application is still at its primitive level. This new emerging technology is far from being mature both in theory and practical implementation [11]. There are a few technologies in the area of DNA research which are accepted in the past decade such as polymerase chain reaction (PCR), DNA digital coding, and DNA synthesis and its implementation in cryptography [12]. Researchers are working in the direction to use the power of DNA computation to strengthen the existing security systems. DNA encryption is hybridized with conventional methods of encryption and some other methods like chaos theory, metaheuristic optimizations which have various applications in different fields [13–24]. It can be useful to secure several types of data. Moreover, some specific regions within an image can be hidden with the help of some interactive environments [25] which is helpful in various domains like military applications and medical applications [26–31].

2 DNA

DNA is the abbreviation for deoxyribonucleic acid which is considered as the basic building block for every living organism. It contains some vital information about life. It constitutes long chains of nucleotides. Nucleotides consist of a base made with nitrogen, carbon, and phosphate. There are four bases that explain and preserve vital information about an organism. These bases are adenine (A), thymine (T), cytosine (C), and guanine (G). These bases are arranged in a double-helical structure that forms a DNA, and it is shown in Fig. 1 [32].

3 DNA Computing

DNA computation is one of the recent trends in computation which is first described in [33]. It has been applied in solving various combinatorial problems. One of the oldest applications of DNA computing was in the Hamiltonian path problem. The digital DNA encoding scheme is given in Fig. 2. This problem was solved by using DNA computation with brute-force method. This approach was modified in [34] and applied on a NP-complete problem. The concept of the above two methods was merged in [35]. It was applied to break the security of one of the most popular cryptographic algorithms called DES. It has been found that the DES method can be broken in near about four months. It explains the power of the DNA computing. In [36], DNA computing is applied to solve the maximal clique problem. It is one of the well-known NP-complete problems. DNA computing proves its efficiency in several instances. It can solve some difficult problems which are not solvable by traditional methods due to its parallelism.

Fig. 1 Structure of DNA

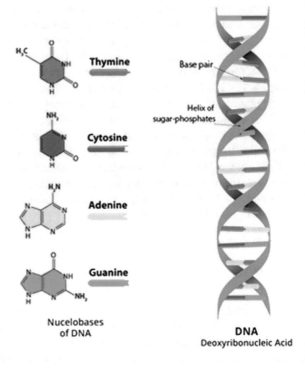

Nucelobases
of DNA

DNA
Deoxyribonucleic Acid

Fig. 2 DNA digital coding
scheme

A	T	G	C
⇩	⇩	⇩	⇩
00	01	10	11

4 Application of DNA Computing in Cryptography

DNA computation is one of the powerful computational tools which is being widely used in different applications. Data security is one of the domains where the DNA computing is frequently used. In this section, a brief review on different cryptographic approaches based on DNA computing is presented. In [37], author proposed a cryptographic method based on DNA computing. In this work, an image encryption method is proposed that uses the molecular behavior of DNA beside some traditional methods like one-time pad. It is a simple approach that was applied on a two-dimensional image. This concept was applied in [38] with some modifications. This method provides high data security. In this work, a DNA chip was used along with a one-time pad. Authors proved that the encoded message was very difficult to find by a third-party entity. In [12], authors proposed a one-way encryption method which is based on DNA cryptography and public key. In this work, author suggested the use of two public keys. Authors have explained the creation process of the two public keys and their application phases. To decode the encoded data, PCR amplification is used with the help of some secret sequences. In [39], authors developed a symmetric key encryption system. In this work, a genetic database is used to retrieve the keys and it remains the same for both the ends. In [40], authors proposed a new data security method based on DNA computing. In this work, pseudo DNA cryptographic method is used in MANET. Its prime objective is to secure the ad hoc networks. These kinds of network suffer from the lack of the centralized control. It is a significant security threat that makes the transmitted data vulnerable. The approach uses pseudo DNA cryptography method which is based on the central dogma principle. The biological process is mimicked to encrypt the data and make the data secure for transmission. Here, one-time pad is used and it is a symmetric encryption process. The DNA synthesis process is used in a public-key cryptography system which is explained in [11]. Besides the DNA synthesis process, the proposed method uses DNA coding and polymerize chain reaction (PCR). This method provides a high degree of security.

5   Literature Review

Although it is not as mature as other fields in cryptography, researchers have done a lot of work instead of its practical limitations. In [37], author has proposed a new DNA-based cryptographic approach to perform encryption and decryption of a two-dimensional image using one-pad technique. Ashish Gehani et al. [38] worked on the molecular theory and established the foundation of DNA cryptography. Authors extend the work of Vernam and Shannon's one-time pad techniques and proved it has perfect secrecy. They argued the present application of one-time pad-based cryptographic system is limited to confines of conventional electronic media, whereas DNA has high information density. Khalifa et al. [41] have shown a method to transform normal message into collection of amino acids and encrypted using DNA-based play-fair cipher. Afterward, to make the quality better, DNA complementary substitution is used. In [42], authors have proposed a new hybridized algorithm to transfer message securely using RSA combined with DNA computing. In [43], authors have demonstrated a new method of encryption based on DNA sequencing. Tanaka et al. [12] have proposed a new approach based on public key using DNA. It is a one-way method. Messages are encoded with DNA sequence with public key and further processed using few processes [44]. Menaka [45] demonstrated a new approach for message encryption using DNA sequencing based on Watson-Crick DNA strand complementary rule. In [46], Kang Ning proposed a symmetric key algorithm which creates fake DNA sequences for cryptographic techniques using complementary rule. It is very useful for network security. Lai et al. [47] proposed a new asymmetric encryption method. It comprises of genetic engineering and cryptography. Najaf et al. [48] proposed an innovative way to make the hybrid cryptography better using DNA steganography and public key. It aims to reduce the usage of public key. Authors also proposed a protocol based upon DNA coding and complementary rules. Verma et al. [40] proposed to secure routing in mobile ad hoc networks that uses pseudo DNA

Table 1 Summary of the works that applies DNA computing in cryptography

References	Brief description about the method
[50]	Exploited the random nature of the DNA and proposes a molecular cryptography scheme based on symmetric key. This work proposes a message transformation method based on carbon nanotube. The proposed method is evaluated using numerical parameters
[51]	This work uses the advantage of DNA computing in terms of parallelism, information density, energy efficiency, etc. This work is a combination of traditional cryptography with DNA computing with symmetric key approaches. PCR amplification and DNA digital coding scheme also embedded in this method. Authors show that the biologically complex problems combined with cryptographic approaches provide two layers of security. The quality of the proposed method is analyzed in details
[34]	This work proposes a new method that combines the DNA method with the one-time pad. Data can be secured in DNA sequences. This article describes both cryptography and steganography
[47]	This work proposes a one-way function with DNA cryptographic system. It is a public-key cryptography system. The target is encrypted inside DNA sequence and can be recovered using PCR amplification. A novel key distribution mechanism is also proposed which is based on public-key system and DNA
[49]	This work proposes a genetic engineering and cryptography-based encryption–decryption mechanism namely DNA-PKC. It is an asymmetric encryption method. The security of the proposed method does not depend on the computational complexity. It exploits the advantages of the biological complexities, and hence, it can prevent quantum computing attacks
[36]	The proposed method combines the strength of the genetic material and the conventional methods. The genomic database is used to access digital DNA sequences. Genomic databases are also helpful in representing one-time pad. Cryptanalysis and statistical evaluation are performed to test the quality of the algorithm

(continued)

Table 1 (continued)

References	Brief description about the method
[52]	This work proposes a hybrid method based on DNA computing. It uses DNA encryption and one-time pad. The time complexity of this algorithm is proven to be better than some other methods
[53]	This work exploits the cryptographic powers of DNA. The original data is transformed into binary sequences which in turn transformed into DNA nucleotides. This algorithm is proven to be robust against different types of cryptographic attacks
[54]	A symmetric key encryption mechanism is proposed with the help of DNA digital coding and DNA fragment assembly. Authors studied the difficulties which are associated with the DNA fragment assembly. Moreover, the properties of the DNA molecules and its applications in the encryption are briefly discussed. Authors prove its high confidential strength by analyzing the security of the method
[55]	In this work, authors propose a new method that combines DNA encryption and chaos theory. Chaotic logistic map is used with the DNA substitution. It is a symmetric key method, and the efficiency of the algorithm is proven using various statistical parameters and visual experiments
[56]	This work uses DNA encryption with bi-objective genetic optimization. The genetic algorithm tries to optimize objective functions namely entropy with correlation coefficient, unified average changing intensity, i.e., UACI, and number of pixel change rate (NPCR). This method improves the performance gain compared to single-objective approach

cryptography. This approach is based on central dogma of molecular theory. Tornea et al. [49] proposed a DNA-based cipher which is based on indexing characteristic of DNA. Authors used a random DNA sequence and use it as one-time pad key which is transferred securely to receiver. The encryption mechanism follows this step by converting a plain text into its equivalent ASCII code and then converted into binary code then it is mapped to DNA sequence and the index number is stored. This array of integers is the ciphertext, and it is decrypted by the receiver with the help of key and index pointer. A brief summary of different works is presented in Table 1.

6 Conclusion

DNA encryption exploits several methods and concepts of biology. It is beneficial in terms of security to exploit the advantages of the difficult biological phenomenon. Different biological concepts and methods like DNA substitution, central dogma, PCR amplification, and DNA synthesis are successfully used with traditional cryptographic techniques. It provides better security that the traditional systems because molecular computations are inherent in it. Hybridization provides better security and makes the method robust against various cryptographic attacks. Moreover, it is helpful in gaining higher confidentiality and information density compared to conventional cryptographic systems. The advantages of this domain attract many researchers to work in this domain, but still some problems like environmental influences and quantum attacks are major challenges that are to be overcome. This work can be certainly helpful in further research in this domain. Several new techniques are evolving and performing better. DNA computation can be hybridized with new methods like chaos theory to make it more robust and resilient against different attacks.

References

1. Sarddar, D., Chakraborty, S., Roy, M.: An efficient approach to calculate dynamic time quantum in round Robin algorithm for efficient load balancing. Int. J. Comput. Appl. **123**(14), 48–52 (2015)
2. Mali, K., Chakraborty, S., Roy, M.: A study on statistical analysis and security evaluation parameters in image encryption. IJSRD Int. J. Sci. Res. Dev. **3**, 2321–2613 (2015)
3. Cui, G., Qin, L., Wang, Y., Zhang, X.: Information security technology based on DNA computing. In: 2007 International Workshop on Anti-Counterfeiting, Security and Identification (ASID), pp. 288–291 (2007)
4. Seal, A., Chakraborty, S., Mali, K.: A new and resilient image encryption technique based on pixel manipulation, value transformation and visual transformation utilizing single–level haar wavelet transform, vol. 458 (2017)
5. Mali, K., Chakraborty, S., Seal, A., Roy, M.: An efficient image cryptographic algorithm based on frequency domain using Haar wavelet transform. Int. J. Secur. Appl. **9**(12), 279–288 (2015)

6. Hore, S., Chatterjee, S., Chakraborty, S., Shaw, R.K.: Analysis of different feature description algorithm in object recognition (2016)
7. Chakraborty, S., Chatterjee, S., Ashour, A.S., Mali, K., Dey, N.: Intelligent computing in medical imaging: a study. In: Dey, N. (ed.) Advancements in applied metaheuristic computing IGI global, pp. 143–163 (2017)
8. Chakraborty, S., Roy, M., Hore, S.: A study on different edge detection techniques in digital image processing. In: Feature Detectors and Motion Detection in Video Processing. IGI Global, pp. 100–122 (2016)
9. Roy, M., Mali, K., Chatterjee, S., Chakraborty, S., Debnath, R., Sen, S.: A study on the applications of the biomedical image encryption methods for secured computer aided diagnostics. In: 2019 Amity International Conference on Artificial Intelligence (AICAI), pp. 881–886 (2019)
10. Basha, D.: Analysis on DNA based cryptography to secure data transmission (2011)
11. Cui, G., Qin, L., Wang, Y., Zhang, X.: An encryption scheme using DNA technology. In: Bio-Inspired Computing: Theories and Applications, pp. 37–42 (2008)
12. Tanaka, K., Okamoto, A., Saito, I.: Public-key system using DNA as a one-way function for key distribution. Biosystems $81(1)$, 25–29 (2005)
13. Chakraborty, S., et al.: Modified cuckoo search algorithm in microscopic image segmentation of hippocampus. Microsc. Res. Tech. $80(10)$, 1051–1072 (2017)
14. Chakraborty, S., Bhowmik, S.: Blending roulette wheel selection with simulated annealing for job shop scheduling problem. In: Michael Faraday IET International Summit 2015, p. 100(7) (2015)
15. Chakraborty, S., Bhowmik, S.: An efficient approach to job shop scheduling problem using simulated annealing. Int. J. Hybrid Inf. Technol. $8(11)$, 273–284 (2015)
16. Chakraborty, S., Bhowmik, S.: Job shop scheduling using simulated annealing. In: First International Conference on Computation and Communication Advancement, vol. 1(1), pp. 69–73 (2013)
17. Chakraborty, S., Seal, A., Roy, M.: An elitist model for obtaining alignment of multiple sequences using genetic algorithm. In: 2nd National Conference NCETAS 2015, vol. 4(9), pp. 61–67 (2015)
18. Roy, M., et al.: Biomedical image enhancement based on modified Cuckoo search and morphology. In: 2017 8th Annual Industrial Automation and Electromechanical Engineering Conference (IEMECON), pp. 230–235 (2017)
19. Chakraborty, S., et al.: Detection of skin disease using metaheuristic supported artificial neural networks. In: 2017 8th Annual Industrial Automation and Electromechanical Engineering Conference (IEMECON), pp. 224–229 (2017)
20. Chakraborty, S., et al.: Image based skin disease detection using hybrid neural network coupled bag-of-features. In: 2017 IEEE 8th Annual Ubiquitous Computing, Electronics and Mobile Communication Conference, UEMCON 2017, vol. 2018 (2018)
21. Chakraborty, S., et al.: Bio-medical image enhancement using hybrid metaheuristic coupled soft computing tools. In: 2017 IEEE 8th Annual Ubiquitous Computing, Electronics and Mobile Communication Conference, UEMCON 2017, vol. 2018 (2018)
22. Chakraborty, S., Mali, K.: Application of multiobjective optimization techniques in biomedical image segmentation—a study. In: Multi-objective optimization. Springer, Singapore, pp. 181–194 (2018)
23. Chakraborty, S., Raman, A., Sen, S., Mali, K., Chatterjee, S., Hachimi, H.: Contrast optimization using Elitist metaheuristic optimization and gradient approximation for biomedical image enhancement. In: 2019 Amity International Conference on Artificial Intelligence (AICAI), pp. 712–717 (2019)
24. Chakraborty, S., Chatterjee, S., Chatterjee, A., Mali, K., Goswami, S., Sen, S.: Automated breast cancer identification by analyzing histology slides using metaheuristic supported supervised classification coupled with bag-of-features. In: 2018 Fourth International Conference on Research in Computational Intelligence and Communication Networks (ICRCICN), pp. 81–86 (2018)

25. Hore, S., et al.: An integrated interactive technique for image segmentation using stack based seeded region growing and thresholding. Int. J. Electr. Comput. Eng. 6(6) (2016)
26. Hore, S., et al.: Finding contours of hippocampus brain cell using microscopic image analysis. J. Adv. Microsc. Res. 10(2), 93–103 (2015)
27. Chakraborty, S., Chatterjee, S., Dey, N., Ashour, A.S., Shi, F.: Gradient approximation in retinal blood vessel segmentation. In: 2017 4th IEEE Uttar Pradesh Section International Conference on Electrical, Computer and Electronics (UPCON), pp. 618–623 (2017)
28. Roy, M., et al.: Cellular image processing using morphological analysis. In: 2017 IEEE 8th annual ubiquitous computing, electronics and mobile communication conference (UEMCON), pp. 237–241 (2017)
29. Chakraborty, S., et al.: Dermatological effect of UV rays owing to ozone layer depletion. In: 2017 4th International Conference on Opto-Electronics and Applied Optics (Optronix), pp. 1–6 (2017)
30. Chakraborty, S., et al.: Bag-of-features based classification of dermoscopic images. In: 2017 4th International Conference on Opto-Electronics and Applied Optics. Optronix 2017, vol. 2018 (2018)
31. Chakraborty, S., et al.: An integrated method for automated biomedical image segmentation. In: 2017 4th International Conference on Opto-Electronics and Applied Optics, Optronix 2017, vol. 2018 (2018)
32. DNA vs RNA—Barca.fontanacountryinn.com: [Online]. Available: http://barca. fontanacountryinn.com/dna-vs-rna/. Accessed: 08 Feb 2019
33. Adleman, L.M.: Molecular computation of solutions to combinatorial problems. Science 266(5187), 1021–1024 (1994)
34. Lipton, R.J.: DNA solution of hard computational problems. Science 268(5210), 542–545 (1995)
35. Boneh, D., Dunworth, C., Lipton, R.J.: Breaking DES using a molecular computer. In: 1st DIMACS workshop on DNA based computers (1995)
36. Ouyang, Q., Kaplan, P.D., Liu, S., Libchaber, A.: DNA solution of the maximal clique problem. Science 278(5337), 446–449 (1997)
37. Chen, J.: A DNA-based, biomolecular cryptography design. In: Circuits System (2003)
38. Gehani, A., LaBean, T., Reif, J.: DNA-based cryptography, pp. 167–188. Springer, Berlin (2003)
39. Amin, S., Saeb, M., El-Gindi, S.: A DNA-based implementation of YAEA encryption algorithm. Comput. Intell. (2006)
40. Verma, A.K., Dave, M., Joshi, R.C.: DNA cryptography: a novel paradigm for secure routing in Mobile Ad hoc Networks (MANETs). J. Discret. Math. Sci. Cryptogr. 11(4), 393–404 (2008)
41. Khalifa, A., Atito, A.: High-capacity DNA-based steganography. In: 8th International Conference on Informatics (2012)
42. Wang, X., Zhang, Q.: DNA computing-based cryptography. In: 2009 Fourth International on Conference on Bio-Inspired Computing, pp. 1–3 (2009)
43. Roy, B., Majumder, A.: An Improved Concept of Cryptography Based on DNA Sequencing (2012)
44. Kaundal, A.K., Verma, A.K.: DNA Based Cryptography: A Review (2014)
45. Menaka, K.: Message encryption using DNA sequences. In: 2014 World Congress on Computing and Communication Technologies, pp. 182–184 (2014)
46. Ning, K.: A Pseudo DNA Cryptography Method. Mar 2009
47. Lai, X., Lu, M., Qin, L., Han, J., Fang, X.: Asymmetric encryption and signature method with DNA technology. Sci. China Inf. Sci. 53(3), 506–514 (2010)
48. Torkaman, M.R.N., Kazazi, N.S., Rouddini, A.: Innovative approach to improve hybrid cryptography by using DNA steganography. Int. J. New Comput. Archit. Appl. 2(1), 225–236 (2012)
49. Tornea, O., Borda, M.E.: Security and complexity of a DNA-based cipher. In: 2013 11th RoEduNet International Conference, pp. 1–5 (2013)

50. Chen, J: A DNA-based, biomolecular cryptography design. In: Proceedings of the 2003 International Symposium on Circuits and Systems, 2003. ISCAS'03, vol. 3, pp. III-822–III-825
51. Cui, G., Qin, L., Wang, Y., Zhang, X.: An encryption scheme using DNA technology. In: 2008 3rd International Conference on Bio-Inspired Computing: Theories and Applications, pp. 37–42 (2008)
52. Pramanik, S., Setua, S.K.: DNA cryptography. In: 2012 7th International Conference on Electrical and Computer Engineering, pp. 551–554 (2012)
53. Amin, S.T., Saeb, M., El-Gindi, S.: A DNA-based implementation of YAEA encryption algorithm. Comput. Intell. (2006)
54. Zhang, Y.: A DNA-based encryption method based on DNA chip and PCR amplification techniques,-9. pdf, works.bepress.com
55. Chakraborty, S., Seal, A., Roy, M., Mali, K.: A novel lossless image encryption method using DNA substitution and chaotic logistic map. Int. J. Secur. Appl. **10**(2), 205–216 (2016)
56. Suri, S., Vijay, R.: A bi-objective genetic algorithm optimization of Chaos-DNA based hybrid approach. J. Intell. Syst. **28**(2), 333–346 (2019)

Author Index

© Springer Nature Singapore Pte Ltd. 2020
M. Chakraborty et al. (eds.), *Proceedings of International
Ethical Hacking Conference 2019*, Advances in Intelligent
Systems and Computing 1065, https://doi.org/10.1007/978-981-15-0361-0